797,885 Books

are available to read at

www.ForgottenBooks.com

Forgotten Books' App
Available for mobile, tablet & eReader

ISBN 978-1-330-08426-7
PIBN 10021749

This book is a reproduction of an important historical work. Forgotten Books uses state-of-the-art technology to digitally reconstruct the work, preserving the original format whilst repairing imperfections present in the aged copy. In rare cases, an imperfection in the original, such as a blemish or missing page, may be replicated in our edition. We do, however, repair the vast majority of imperfections successfully; any imperfections that remain are intentionally left to preserve the state of such historical works.

Forgotten Books is a registered trademark of FB &c Ltd.
Copyright © 2015 FB &c Ltd.
FB &c Ltd, Dalton House, 60 Windsor Avenue, London, SW19 2RR.
Company number 08720141. Registered in England and Wales.

For support please visit www.forgottenbooks.com

1 MONTH OF
FREE
READING
at
www.ForgottenBooks.com

By purchasing this book you are eligible for one month membership to ForgottenBooks.com, giving you unlimited access to our entire collection of over 700,000 titles via our web site and mobile apps.

To claim your free month visit:

www.forgottenbooks.com/free21749

* Offer is valid for 45 days from date of purchase. Terms and conditions apply.

Similar Books Are Available from

www.forgottenbooks.com

A History of Mathematics
by Florian Cajori

Differential Calculus for Beginners
by Joseph Edwards

The Calculus for Engineers
by John Perry

Integral Calculus for Beginners
With An Introduction to the Study of Differential Equations, by Joseph Edwards

On the Definition of the Sum of a Divergent Series, Vol. 1
by Louis Lazarus Silverman

A Treatise on Bessel Functions
And Their Applications to Physics, by Andrew Gray and G. B. Mathews

A Primer of Calculus
by Arthur Stafford Hathaway

Smithsonian Mathematical Formulae and Tables of Elliptic Functions
by Edwin P. Adams

A Treatise on Ordinary and Partial Differential Equations
by William Woolsey Johnson

Smithsonian Mathematical Tables
Hyperbolic Functions, by George F. Becker

The Mathematical Writings of Duncan Farquharson Gregory
by Duncan Farquharson Gregory

Mathematical Handbook
by Edwin P. Seaver

Elements of the Theory of Functions of a Complex Variable
With Especial Reference to the Methods of Riemann, by H. Durège

Vector Calculus
With Applications to Physics, by James Byrnie Shaw

Handbook of Engineering Mathematics
by Walter E. Wynne

Numerical Solution of the Boltzmann Equation
by Alexandre Joel Chorin

Engineering Applications of Higher Mathematics, Vol. 1
Problems on Machine Design, by Vladimir Karapetoff

Calculus
by Herman W. March

Vector Analysis
by Edwin Bidwell Wilson

Introduction to the Calculus of Variations
by William Elwood Byerly

C. H. S. Leicester

LIBRARY OF THE UNIVERSITY OF MICHIGAN

PRINCIPLES

OF THE

DIFFERENTIAL AND INTEGRAL CALCULUS.

LONDON
PRINTED BY SAMUEL BENTLEY,
Dorset Street, Fleet Street.

PRINCIPLES

OF THE

DIFFERENTIAL AND INTEGRAL

CALCULUS,

FAMILIARLY ILLUSTRATED, AND APPLIED TO A

VARIETY OF USEFUL PURPOSES.

DESIGNED FOR THE INSTRUCTION OF YOUTH.

BY THE

REV. WILLIAM RITCHIE, LL.D. F.R.S.

PROFESSOR OF NATURAL PHILOSOPHY AT THE ROYAL INSTITUTION OF GREAT BRITAIN, AND PROFESSOR OF NATURAL PHILOSOPHY AND ASTRONOMY IN THE UNIVERSITY OF LONDON.

J'ai toujours été persuadé qu'un livre élémentaire ne peut être jugé que par l'expérience: qu'il faut, pour ainsi dire, l'essayer sur l'esprit des élèves; et vérifier, par cette épreuve, la bonté des méthodes que l'on a choisies.—Biot, *Essais de Géométrie.*

LONDON:

PRINTED FOR JOHN TAYLOR,

BOOKSELLER AND PUBLISHER TO THE UNIVERSITY OF LONDON,

UPPER GOWER STREET.

1836.

It is somewhat remarkable that, whilst almost every department of human knowledge has been simplified and brought down to the level of ordinary capacities, scarcely any attempt has been made to simplify and illustrate by familiar examples one of the most elegant and useful branches of mathematical science. Most of the existing works on the Differential and Integral Calculus are written for the use of "Students in the Universities," and require a previous knowledge of almost every branch of Pure Mathematics. It is quite obvious, however, that the great leading principles of this science may be communicated to youth at a much earlier period, and with much less acquaintance with the other branches of mathematics than is generally supposed. The pupil may begin to study the Differential and Integral Calculus after he has acquired the elements of Geometry, and the principles of Algebra as far as the end of quadratic equations. Instead of continuing to prosecute his algebraic studies through the Theory of Equations, Indeterminate Problems, Diophantine Analysis, &c. he might advantageously at this period be made acquainted with the simpler applications of Algebra to Geometry, the Elements of Plane Trigonometry, and the Principles of the Calculus. This will be found to be not only a simpler course, but the one that will be most interesting to the pupil, and most useful in expanding his understanding by a new mode of thinking and reasoning. The Fluxionary or Differential and Integral Calculus has within these few years become almost entirely a science of symbols and mere algebraic formulæ, with scarcely any illustration or practical application. Clothed as it is in a transcendental dress, the ordinary student is afraid to approach it; and even many of those whose resources allow them to repair to the Universities do not appear to derive all the advantages which might be expected from the study of this interesting branch of mathematical science. Professor Airy, in his elegant work on "Gravitation," remarks, "It is known to every one who has been engaged in the instruction of students at our universities, that the results of the Differential Calculus are received by many rather with the doubts of imperfect faith than with the confidence of rational conviction." This in all probability arises, not so much from the nature of the subject itself, as from the abstract analytical form in which it is presented to the mind, and the total want of geometrical, arithmetical, and familiar illustration.

Most writers on this subject seem to have paid too little attention to the distinction between the mere analytical operations and the philosophy of the science, and hence the examples and exercises given at the very commencement are as difficult as those given in the more advanced parts of the subject. "It is to be regretted," says Mr. Woolhouse, in his neat and comprehensive Essay on the Principles of the Calculus, "that most of our academical treatises on this as well as other subjects, abound so much with complex algebraic processes, without the slightest traces of logical reasoning to exercise and improve the intellect. We should bear in mind that the simple execution of analytical operations acquired by dint of practice and experience, is a mere common species of labour, often merely mechanical, whilst a distinct apprehension of the specific object and meaning of the operations, and a contemplation of the clearness and beauty of the various arguments employed, constitute the intellectual lore that gratifies and enriches the mind, and stimulates its energies with an ardour after the investigation of truth."

The plan adopted in the present work is founded on the same process of thought by which we arrive at actual discovery, namely, *by proceeding step by step from the simplest particular examples till the principle unfolds itself in all its generality.* Again, the Differential Calculus is generally taught as one science, and the Integral as another; whereas, according to the plan here adopted, they are made to travel hand-in-hand till they arrive at the point where the former naturally stops and the latter advances through fertile regions without requiring further aid from its companion. This plan embraces perhaps as much matter as it will be found practicable to teach in our schools and academies, and in the elementary classes in many of our colleges and universities.

As this little volume is expressly designed for the use of young persons, I have preferred a simple and even familiar style to the formal and more dignified language used in works of higher pretension. If I have succeeded in leading the pupil by easier gradations than other writers, from the mere elements of geometry and algebra to a knowledge of the fundamental principles of the most important branch of mathematical science, I shall have succeeded in accomplishing the task I imposed on myself; and even if I have not succeeded to the extent I could have wished, the work will nevertheless, I trust, be useful to an extensive class of students, who are anxious to know the leading principles of the science, but who find themselves unable to proceed beyond the first few pages of the ordinary treatises on the subject.

London, 10*th Dec.* 1835.

CONTENTS.

PART I.

CONTAINING PRINCIPLES, RULES, AND EXERCISES.

PART II.

APPLICATIONS OF THE PRECEDING RULES AND PRINCIPLES TO USEFUL PURPOSES.

PART III.

DEVELOPEMENT OF ALGEBRAIC EXPRESSIONS INTO INFINITE SERIES, DIFFERENTIATION OF TRANSCENDENTAL FUNCTIONS, AND INTEGRATION BY LOGARITHMS AND ARCS OF CIRCLES.

PART IV.

APPLICATION OF THE PRECEDING PRINCIPLES TO DETERMINE THE RADIUS OF CURVATURE, NATURE OF EVOLUTES, &c. OF THE MORE USEFUL CURVES OF THE SECOND AND HIGHER ORDERS.

APPENDIX.

ERRATA.

Page 11, line 8 from bottom, for $\sqrt{a^2+u^2}$ read $\sqrt{a^2+x^2}$.

12, line 15, for *increment* read *quantity*.

line 25, for *inches* read *feet*.

13, line 29, for *increment* read *quantity*.

19, lines 18 and 19, for axh read $2axh$.

21, line 15, for $\frac{1}{2}x^{-\frac{1}{2}-1}dx$ read $\frac{1}{2}x^{\frac{1}{2}-1}$.

36, line 10, for $\frac{1}{2}ax+bnx$ read $\frac{1}{2}an+bnx$.

39, line 23, for $\frac{b^2-a^2}{2n}$ read $\frac{n(b^2-a^2)}{2}$.

51, line 11, for $(a-x)$ read $(a-x)x$.

line 13, for $2xda$ read $2xdx$.

53, line 4, for $4x^2(ax+x^2)$ read $4x(ax-x^2)$.

53, line 24, for $\pi+\frac{2c}{x}$ read $\pi x^2+\frac{2c}{x}$

60, lines 27 and 28, erase $+y^2$ from the first side of equations.

61, do the same in line 1.

64, line 12, for $\frac{b^2}{a^2}(ax^2-x^2)$ read $\frac{b^2}{a^2}(ax-x^2)$.

line 13, for $\frac{b^2}{a^2}(ax^2+x^2)$ read $\frac{b^2}{a^2}(ax+x^2)$.

65, line 1, for $\frac{b^2}{a}$ read $\frac{b^2}{a^2}$.

72, eq. (1), for $\frac{dx}{\sqrt{2-x^2}}$ read $\frac{dx}{\sqrt{2x-x^2}}$.

73, line 17, for $\frac{t^2}{3}$ read $\frac{t^3}{3}$.

76, line 16, for $\int\frac{2ydy}{p}$ read $\int\frac{2y^2dy}{p}$.

82, line 3, for $8rxdx$ read $8arxdx$.

87, line 1, for $a-2x$ read $2a-2x$.

170, line 6, for *the number of terms* read *half the number of terms*.

PRINCIPLES

OF THE

DIFFERENTIAL AND INTEGRAL CALCULUS.

PART I.

CONTAINING PRINCIPLES, RULES, AND EXERCISES.

INTRODUCTION.

(1.) In the solution of problems by Algebra certain quantities are given, and it is required to determine the value of others from the relation they bear to those that are known. These quantities, whether known or unknown, are supposed to retain their values during the whole process of calculation. In the solution of problems, or other investigations, by means of the Differential Calculus, there are certain quantities which always retain the same values, whilst others are supposed to increase or decrease at a uniform rate. The former are called *Constant Quantities*, and are represented by the first letters of the alphabet, a, b, c, &c.; the latter are called *Variable Quantities*, and are represented by the last letters x, y, z.

(2.) Variable quantities are generally confined within certain *limits*, while others, whose value depends on these and constant quantities, have different limits, which are to be ascertained.

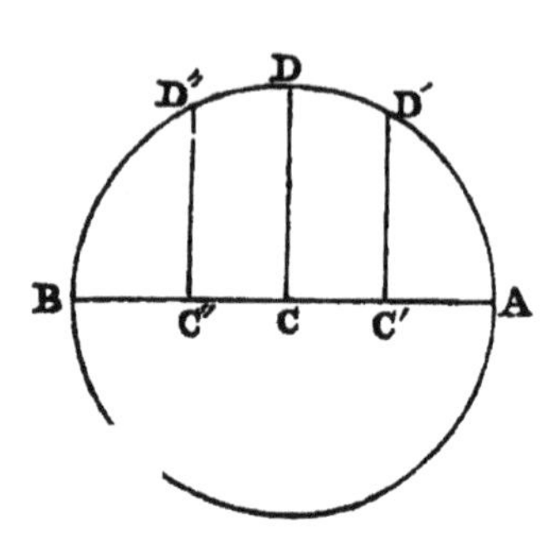

Let A B be the diameter of a circle, and let a line begin to move from A uniformly along A B, its extremity, keeping constantly in the circumference of the circle, then AC′ and C′D′ are *variable* quantities, whilst the diameter of the circle is a *constant* quantity. When the perpendicular has reached the centre of the circle it is equal to the radius, after which it continually diminishes till it become 0 at the point B. The radius of the circle is therefore the limit of the perpendicular.

(3.) If a regular polygon be inscribed in a circle, and if we inscribe another, having twice the number of sides, the area of the second will approach more nearly to the area of the circle than that of the first. By continuing to double the number of sides, the area of the polygon will approach nearer and nearer to that of the circle, and may be made to differ from it by a quantity *less* than any finite quantity. Hence the circle is said to be the limit of the inscribed polygon.

(4.) If we convert $\frac{1}{9}$ into a decimal fraction, it becomes .1111, &c. or $\frac{1}{10} + \frac{1}{100} + \frac{1}{1000}, + $ &c. Hence the series approaches to the value of $\frac{1}{9}$, but can never equal it, whilst the number of terms is finite. We therefore say, $\frac{1}{9}$ is the limit of the series. When the number of terms is infinite, the sum of the series is $\frac{1}{9}$.

(5.) If a number be multiplied by another, the product becomes less and less as one of the numbers diminishes; hence $a \times 0 = 0$, or 0 is the limit of ax when x becomes 0.

(6.) Every teacher of the mathematics must have felt the difficulty of making pupils understand what we really mean when we say, that 1 divided by 0 is infinite. Now, the best mode of teaching is to make the pupil discover for himself by a series of leading questions, such as the following:—

How much is 1 divided by 1? Ans.: 1.

How much is 1 divided by $\frac{1}{10}$? Ans.: 10.

How much is 1 divided by a fraction, having 1 for its numerator, and 1 with as many ciphers as would reach to one of the fixed stars for its denominator? Ans.: 10,000,000, &c. to the fixed stars.

Hence, as the divisor approaches 0, what does the quotient approach? Ans.: Infinity.

Hence, by an extension of reasoning, when the divisor becomes nothing, what is the quotion or value of the fraction? Ans.: Infinite.

(7.) After this familiar illustration the teacher may prove the property more scientifically thus:—

$$\frac{1}{1-x} = 1 + x + x^2 + x^3 +, \text{ and to infinity.}$$

When $x=1$, $\frac{1}{0} = 1 + 1 + 1 + 1 +$, &c. to infinity, which is an infinite quantity.

(8.) Since the value of a fraction diminishes by increasing the denominator, it follows, that when the denominator becomes infinite, the value of the fraction, or its limit, is 0.

(9.) As the denominator of a fraction approaches to the value of the numerator, the value of the fraction approaches 1, and when it becomes equal the value is 1. Now $\frac{a^m}{a^n} = a^{m-n}$, and when $m = n$, $\frac{a^m}{a^m} = a^0 = 1$.

Hence any quantity raised to the power 0 is 1.

The pupil is required to illustrate this property as in the preceding examples.

(10.) To find the limit of $\frac{a^2-b^2}{a-b}$ when b approaches to the value of a, and ultimately becomes equal to it.

By actual division $(a^2-b^2) \div (a-b) = a+b$, and when $a = b$, the value or limit is $2a$.

NUMERICAL ILLUSTRATION.

Let $a = 3$ and $b = 2$, then $\dfrac{3^2-2^2}{3-2} = 5$.

Again, let $a = 10$ and $b = 9$, then $\dfrac{10^2-9^2}{10-9} = 19$.

In the first example, 5 approaches to $2a$ or 6, and differs from it only by $\frac{1}{5}$ of 5. In the second example 19 differs from $2a$ or 20 only by the $\frac{1}{19}$ part of itself. Hence, as the numbers approach more nearly to a *ratio* of equality, the value of the expression approaches more nearly to $2a$.

This will be rendered clearer if we convert the terms into proportional quantities; thus,

$\dfrac{a^2-b^2}{a-b} = \dfrac{a+b}{1}$ or $a^2-b^2 : a-b :: a+b : 1 :: 2a : 1$

when $a = b$. That is, the ratio between the difference of the squares of two numbers, and the difference of these numbers, approaches nearer and nearer to the ratio of twice the greatest to unity, as these numbers approach to a ratio of equality.

(11.) What is the value of the fraction $\dfrac{2x+5}{4x+6}$ when x becomes infinite?

Dividing both numerator and denominator by x we have $\dfrac{2+\frac{5}{x}}{4+\frac{6}{x}}$; but when x becomes infinite, the fractions $\dfrac{5}{x}$ and $\dfrac{6}{x}$ become nothing.

Hence the limit is $\frac{2}{4}$ or $\frac{1}{2}$.

(12.) What is the limit to which the ratio of h to $2xh + h^2$ approaches as h diminishes, and ultimately becomes equal to 0?

$h : 2xh + h^2 :: 1 : 2x + h$, and when $h = 0$, the ratio is that of 1 to $2x$.

(13.) To determine an expression for the tangent of a circle.

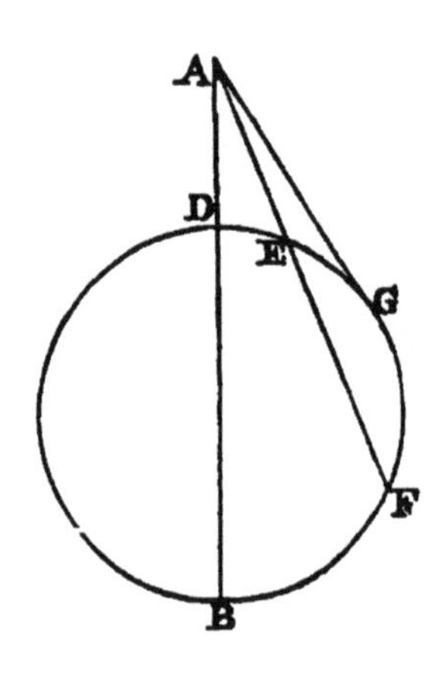

Let AB, AE be lines drawn from the point A, cutting the circle. Then AB $\times$ AD $=$ AE $\times$ AF [*]. Let AE $= x$ and EF $= h$, then AF $= x + h$. And AB $\times$ AD $= (x + h)\,x - x^2 + xh = x^2 =$ AG², when $h = 0$. Hence by this application of the theory of limits we demonstrate the property that the rectangle contained by AB and AD is equal to the square of the tangent, which the pupil may see demonstrated in a purely geometrical manner in "Euclid's Elements."

From these examples and illustrations the pupil will perceive that by the term *limit*, we mean a fixed quantity to which a variable quantity continually approaches, by making certain suppositions, and which may be made to approach nearer to it than by any finite quantity, however small. These suppositions are, generally, that one of the quantities becomes

[*] Euc. III. 36, or Principles of Geometry, iv. Prop. 2.

indefinitely great or indefinitely small, that is, becomes infinite or 0, or, that some of the variables become equal to constant quantities.

(14.) The intelligent pupil may now perhaps ask, how it happens that, since $\frac{a^2-b^2}{a-b} = a+b$, the first side of the equation becomes $\frac{0}{0}$, and the second $2a$; or, in other words, $\frac{0}{0} = 2a$? The question is a very important one, and must be answered to the satisfaction of the pupil. When we speak of the limit of $\frac{a^2-b^2}{a-b}$ when $a=b$, we do not mean the *limit of the numerator* divided by the *limit of the denominator;* but we mean the *limit of the quotient* resulting from actually dividing the numerator by the denominator, which, when $a=b$, is $2a$. In the former case we often arrive at an expression $\frac{0}{0}$, which viewed by itself is absolutely unintelligible.

If we make $b = a-h$ the expression $\frac{a^2-b^2}{a-b}$ becomes $\frac{a^2-(a-h)^2}{a-(a-h)} = \frac{a^2-a^2+2ah-h^2}{h} = \frac{2ah-h^2}{h}$.

Now, if we make $h=0$, which is the same thing as making $a=b$, the limit of the numerator divided by the limit of the denominator is $\frac{0}{0}$; but this is not the limit we wish to find, that being the limit of the quotient resulting from actually dividing the numerator by the denominator. Now if we actually divide by h the quotient is $2a-h$, and when $h=0$ it becomes $2a$, the true limit, as before.

Hence, the equation $\frac{0}{0} = 2x = 3x^2$, &c. deduced from taking the term *limit* in two different senses, when 0 means *absolutely* nothing, is unintelligible, and is one of the stumbling-blocks which has been unnecessarily thrown in the way of the pupil, and that too in the very gate which leads to one of the most fertile domains of science.

EXERCISES.

1. What is the limit of $\frac{a^3-b^3}{a-b}$ when $a = b$?

2. What is the limit of $\frac{x+1}{2x+1}$ when x becomes infinite?

3. What is the limit of the ratio $\frac{xh+h^2}{h}$ when $h=0$?

4. What is the limit to which the ratio of h^2 to $3x^2h^2 + 3xh^3 + h^4$ approaches as h diminishes, and ultimately vanishes?

SECTION I.

ON THE FUNDAMENTAL PRINCIPLE OF THE DIFFERENTIAL CALCULUS, WITH ILLUSTRATIONS.

(1.) If a variable quantity increase uniformly, then other quantities depending on this and constant quantities, may either increase uniformly or according to any variable law whatever.

1. Thus, if a variable quantity, x, increase uniformly, then $2x$ or $3x$, or any given number of times x, will also increase uniformly. Let x increase uniformly by the quantities 1, 1, 1, &c. then its successive values will be $x + 1$, $x+2$, $x + 3$, &c.; hence if we take ax, its corresponding values will be $ax + a$, $ax + 2a$, $ax + 3a$, &c. which increase uniformly by the quantity a. It is also obvious, that if a constant quantity be added to ax the new value will also go on increasing uniformly. Thus, if we take the expression $ax + b$, its successive values will be $ax + a + b$, $ax + 2a + b$, $ax + 3a + b$; which obviously go on increasing uniformly at the same rate with ax.

GEOMETRICAL ILLUSTRATION.

Let ABC be a right-angled triangle, and let a variable line, DE, at right angles to AB, move uniformly along AB, its other extremity, remaining in AC, then it is obvious this line will also increase uni-

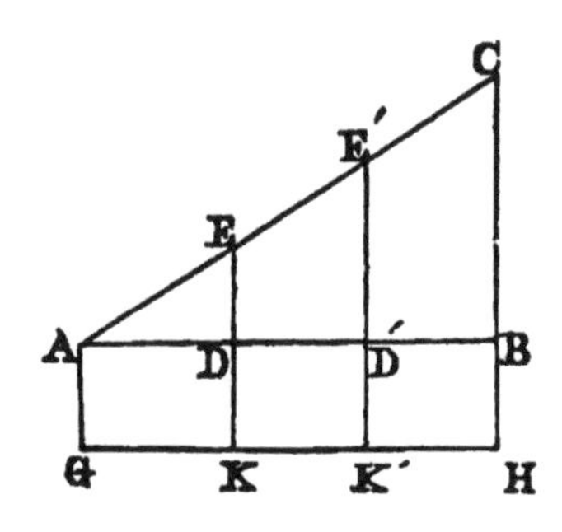

formly, for in every position we have AD : DE :: AD′ : DE′.

Again, let EK, which is equal to ED + DK, be the new variable line, then, whilst it moves uniformly along GH, it will increase at the same uniform rate as before.

FAMILIAR ILLUSTRATION.

3. Take a small thread of indian-rubber, and fix it between two points; let it be moved uniformly along AB, and at right angles to it, one of the points being kept on AB, and the other on AC, and it will increase at a uniform rate.

(2.) It most frequently happens that the quantities we have to consider do not increase uniformly, whilst the independent variable increases at a uniform rate. If, for example, x increase uniformly, x^2 does not increase uniformly.

For let x increase uniformly by the quantities 1, 1, 1, so as to become successively $x + 1$, $x + 2$, $x + 3$, then the values of the squares of these quantities do not increase uniformly. By squaring these quantities the successive values will be $x^2 + 2x + 1$, $x^2 + 4x + 4$, $x^2 + 6x + 9$, in which the successive differences are $2x + 1$, $2x + 4$, $2x + 9$, which are not equal, but go on constantly increasing.

GEOMETRICAL ILLUSTRATION.

Let AB be the side of a square, and let it increase uniformly by the increments, 1, 2, 3, so as to become AB + 1, AB + 2, AB + 3, &c. and let squares be described on the new sides,

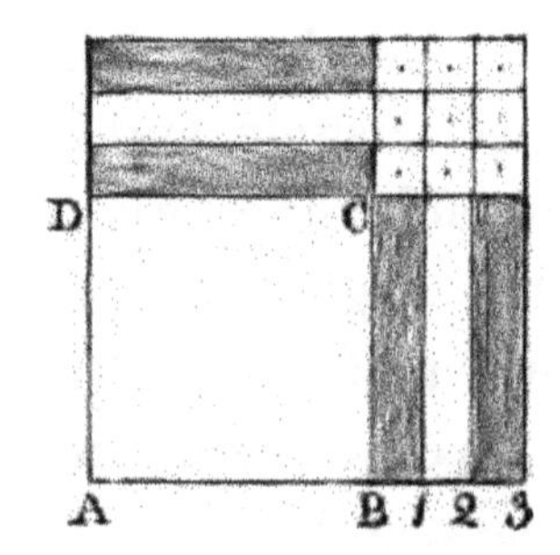

as in the annexed figure; then it is obvious that the square on the side A 1 exceeds that on AB by the two shaded rectangles and the small white square in the corner. The square described on A2 has received an increase of two equal rectangles with *three* equal white squares in the corner. The square on A3 has received an increase of two equal rectangles and *five* equal small squares. Hence, when the side increases uniformly the area goes on at an increasing rate.

(3.) When the value of a quantity depends upon the particular value of another variable quantity, the one quantity is said to be a *function* of the other.

Thus, the area of a square, depending on the length of its side, is said to be *a function of the side*. The algebraic expressions ax^2, $\sqrt{a^2+x^2}$, $\frac{a+bx^3}{ax}$, &c. depending for their value on that which we assign to x, are all functions of x. By the term *function of* x, then, we mean any algebraic expression into which x enters in combination with constant quantities. If the expression be put equal to a single letter, thus, $u = \sqrt{a^2+x^2}$, x is called the *independent variable*, and u, or its equal $\sqrt{a^2+x^2}$, the *dependent variable*.

(4.) The object of the differential calculus, is to determine the *ratio* between the rate of variation of the independent variable and that of the function into which it enters.

Ex. If the side of a square increase uniformly, at what rate does the area increase when the side becomes x?

Let x become $x+h$, then the new area is $(x+h)^2$, or $x^2 + 2xh + h^2$. Whilst the side increases by h the area has increased by $2xh + h^2$. If h denote the rate at which x increases, then $2xh + h^2$ would have denoted the rate at which the area increases had that rate been uniform. In that case we should have had the following proportion:—Rate of increase of the side : rate of increase of the area :: $h : 2xh + h^2$, or dividing both terms by h, as $1 : 2x + h$. But as the area of the square goes on increasing more rapidly than the side, the quantity $2x + h$ is greater than the increment, which the area would have received, had the rate at which the area was increasing continued uniform. Now, the smaller h becomes the nearer does the increment $2x+h$ approach to that which would have resulted had the rate at which the area was increasing when the side became x continued uniform. Hence, when $h = 0$ the ratio becomes that of 1 to $2x$.

NUMERICAL ILLUSTRATION.

If the side of a square increase uniformly at the rate of three feet per second, at what rate is the area increasing when the side becomes 10 feet? $1 : 2x :: 3 : 6x$. Hence the rate of increase is 6×10, or at the rate of 60 square inches per second.

DEFINITION 1. The uniform rate at which the independent variable increases was by Newton called its *fluxion*, and was represented by a point or dot placed above it; and the rate at which the function varied at the same instant, its *fluxion*.

Thus, if $u = x^2$, then, according to Newton's notation,

$$\dot{x} : \dot{u} :: 1 : 2x.$$

$$\text{and } \dot{u} = 2x\dot{x}$$

According to the views and notation of Leibnitz, the rate at which a quantity varies uniformly was called its *differential*, and denoted by the letter d placed before it, thus dx. Hence the proportion is expressed by $dx : du :: 1 : 2x$.

$$\text{And } du = 2xdx.$$

The differential of a compound quantity is expressed by inclosing the expression by a parenthesis and placing $d.$ before it; thus, $d.(a+x^2) = 2xdx$.

Also $d.\,xy$ means the differential of the product of xy.

DEF. 2. The quantity by which dx is multiplied, so as to render it equal to du, is called the *Differential Coefficient;* thus $2x$ is the differential coefficient.

The differential of a constant quantity is 0.

(5.) If the side of a cube vary uniformly, at what rate does the solidity vary when the side becomes x?

Let $h =$ the increment of the side, then the expression for the solidity of the new cube will be $(x+h)^3$, or $x^3 + 3x^2h + 3xh^2 + h^3$, and its increment above x^3 is $3x^2h + 3xh^2 + h^3$. Now, had the solidity of the cube varied uniformly we should have had the increments proportional to the rates of increase; or, rate of increase of the side : rate of increase of the solidity $:: h : 3x^2h + 3xh^2 + h^3$, or as 1 to $3x^2 + 3xh + h^2$; but as the solidity goes on increasing much more rapidly than that of the side, the increment $3x^2 + 3xh + h^2$ is greater than that which would have resulted had the rate at which the cube was increasing, when its side became x, continued uniform. Now the smaller h becomes, the nearer will this expression approach to $3x^2$, and when $h = 0$ the ratio becomes 1 to $3x^2$. Hence, if the solidity of the cube whose side is x be represented by u or $u = x^3$, then

$$dx : du :: 1 : 3x^2.$$

Hence $du = 3x^2dx$.

NUMERICAL ILLUSTRATION.

At what rate is the solidity of a cube increasing when its side increases at the rate of one inch per second, at the moment its side becomes 10 inches?

$$dx : du :: 1 : 3 \times 10^2, \text{ or } 1 \text{ to } 300.$$

Hence, if the rate at which it was increasing at the instant its side became 10 inches had been continued uniform during the next second, it would have received an increase of 300 cubic inches.

(6.) If instead of x^2, or x^3, we had employed the expressions $x^2 + c$ or $x^3 + c$, and determined the ratio of the rate of increase of x to that of $x^2 + c$, or $x^3 + c$, or in short, of any expression having a constant quantity connected with it by the signs *plus* or *minus*, the ratio would have been found to be the same as for x^2 or x^3, the constant always disappearing when the former state of the expression is taken from the latter. The pupil may go through the operation by substituting $x + h$ for x.

(7.) Though we have employed these palpable illustrations to give the pupil a clear view of the subject, he may now view a function or algebraic expression without any reference to geometrical or other quantities, and determine an expression for the rate of increase of the abstract expression, compared with the uniform rate of the independent variable. He has only to substitute $x + h$ for x, and subtract the former function from the new value, and find the limit to which the ratio approaches when $h = 0$.

EXERCISES.

1. If x increase uniformly at the rate of 1, at what rate does the value of the expression $a + 2x^2$ increase when $a = 4$ and $x = 6$?

2. If x increase uniformly at the rate of 2, at what rate does ax^2 increase when $a = 4$ and $x = 10$?

3. If x increase uniformly at the rate of .1 per second, at what rate does $\frac{x^2}{a}$ increase when x becomes 4, the constant, a being equal to 10?

SECTION II.

NATURE AND OBJECT OF THE INTEGRAL CALCULUS ILLUSTRATED.

(1.) The Integral Calculus is the reverse of the Differential Calculus, its object being to determine the expression or function from which a given differential has been derived. Thus we have found that the differential of x^2 is $2xdx$; therefore, if we have given $2xdx$ we know that it must have been derived from x^2 or $x^2 + C$. The function from which the given differential has been derived is called its *Integral*. Hence, as we are not certain whether the integral has a constant quantity or not added to it, we add a constant quantity denoted by C, the value of which is to be determined from the nature of the problem.

(2.) The view which we have given of the nature of an integral is true as far as it goes; but it frequently happens that we have given a particular form of a differential, which cannot be derived from any algebraic expression. All we can do in such a case is, to find by approximation the nearest value we can obtain.

(3.) Leibnitz considered the differentials of functions as indefinitely small differences, and the *sum* of these indefinitely small differences as making up the function; hence the letter s was placed before the differential to show that the sum was to be taken. As *s* had often to be placed before a compound expression it was elongated into the sign $\int$, which being placed for a differential, denotes that its integral is to be taken. Thus,

$$\int dx = x + C, \int 2x dx = x^2 + C, \int 3x^2 dx = x^3 + C.$$

The sign $\int_x$ has lately been introduced to denote the integral of a differential whose variable is x; thus $\int_x, = x + C$, $\int_x 2x = x^2 + C$, $\int_x 3x = x^3 + C$. We shall use the former notation as being that most generally employed.

(4.) Since a constant multiplier or divisor in a function is a constant multiplier or divisor in the differential, it follows, that the integral of any differential multiplied by a constant quantity is the same as the integral of the other part multiplied by the constant. Thus, since the differential of ax is adx, or $d.ax = adx$, it follows that $\int adx = a\int dx$. Hence, a constant multiplier, whether whole or fractional, following the sign $\int$ may be removed and placed before it.

(5.) If we ask a young pupil to tell us in plain language, divested of technical terms, what we really mean when we say we have given a certain differential to find its integral, he will, generally speaking, find himself very much at a loss for a satisfactory answer. In plain language, then, we have given a quantity which varies uniformly, and the ratio of its rate of variation with *another quantity* depending on it and given quantities, to find the value of that *quantity*. Thus, for example, if we have given the differential $3x^2dx$ to find its integral, we have given a quantity x, which varies uniformly, and the ratio of its variation to that of the quantity which we have to find, viz. the ratio of 1 to $3x^2$, to find the quantity. Since $\int 3x^2dx = x^3$, we have x^3, the quantity required.

NUMERICAL EXAMPLE.

There is a quantity, x, which increases uniformly at the rate of 2 per second, and the rate of its variation compared with another quantity depending on it, is as 1 to ax^2; required the value of this quantity when $a = 12$ and $x = 10$.

Let $u =$ quantity required; then $dx : du :: 1 : ax^2$.
Hence $du = ax^2dx$.

And $\int du$, or $u - \int ax^2dx = \frac{ax^3}{3}$. Hence,

$\frac{12}{3} \times 10^3 \times 2 = 8000$, the number required.

SECTION III.

INVESTIGATION OF THE RULES FOR DIFFERENTIATING AND INTEGRATING THE SIMPLER FORMS OF FUNCTIONS AND DIFFERENTIALS.

(1.) If we have any two algebraic expressions each of which contains x and known quantities, and if the value of these expressions be always equal to each other whatever value we assign to x, then the differential of these expressions must be equal to one another.

This is obvious, for since these quantities are always equal, they vary at the same rate, or their differentials are equal.

(2.) If the independent variable or its function be multiplied or divided by a constant number, that number or quantity remains as a multiplier or divisor in the differential; but if it be connected by the signs *plus* or *minus* it disappears.

Ex. Find the differential of $ax^2 + C$.

Substitute $x + h$ for x, then $a(x + h)^2 + c$ becomes $ax^2 + 2axh + ah^2 + c$, and subtracting the primitive function, the increment is $axh + ah^2$. Hence dx : differential of $ax^2 + c$ :: h : $axh + ah^2$, or, as 1 to $2ax + ah$, provided the function had increased uniformly, or taking the limit when $h = 0$, the ratio becomes that of 1 to $2ax$. Hence,

Differential of $ax^2 + c$ is $2axdx$.

(3.) To find a rule for differentiating x^2 or x^3.

If x become $x + h$, then by taking the second or third power of $x + h$, the first term will be x^2 or x^3, and the powers of x will go on diminishing by 1 till x disappear from the result. The coefficient of the second term is the same as the index of the power, and also contains h as a multiplier. Hence, when the increment is divided by h, this quantity entirely disappears from the second term, but not from the others, so that this term remains when the limit is taken, whilst all the subsequent terms disappear.

Hence, if $u = x^2$, $\frac{du}{dx} = 2x$, or $du = 2xdx$.

If $u = x^3$, $\frac{du}{dx} = 3x^2$, or $du = 3x^2dx$.

RULE. Multiply by the index, diminish the index by unity, and multiply by the differential of the variable.

EXERCISES.

1. What is the differential of $4ax^3$?
2. What is the differential of $\frac{2}{3}x^2 + b$?
3. What is the differential coefficient of $\frac{3x^3 - a^2}{b}$?

4. If x increase uniformly at the rate of 1 per second, at what rate is the expression $\frac{4x^3 + a}{b}$ increasing when x becomes 10, a being $= 4$ and $b = 6$?

(4.) To find the integral of a differential of the form xdx or x^2dx.

Rule. Add one to the index, divide by the index thus increased, and by the differential of the variable.

Ex. $\int 2xdx = \frac{2x^2dx}{2dx} = x^2 + C.$

EXERCISES.

1. What is the integral of xdx?

2. What is the integral of $\frac{x^2dx}{3}$?

3. The rate of variation of the independent variable x, is to the rate of variation of a certain algebraic expression, as 1 to $\frac{a}{b}x^2$; it is required to find that expression.

(5.) To find the differential of $\sqrt{x}$ or $\sqrt[3]{x}$.

Let $u = \sqrt{x} = x^{\frac{1}{2}}$, then $u^2 - x$.

And $2udu = dx$, therefore $du = \frac{dx}{2u}$.

Hence $du - \frac{dx}{2x^{\frac{1}{2}}} = \frac{1}{2} x^{-\frac{1}{2}} dx = \frac{1}{2} x^{\frac{1}{2}-1} dx.$

In like manner it may be shown, that if $u - \sqrt[3]{x}$

$$du = \frac{1}{3} x^{\frac{1}{3}-1} dx = \frac{1}{3} x^{-\frac{2}{3}} dx.$$

Rule. Multiply by the fractional index, diminish the index by unity, and multiply by the differential of the variable.

EXERCISES.

1. What is the differential of $2\sqrt{x}$?

2. What is the differential of $\frac{a}{b}\sqrt[3]{x}$?

3. If the *area* of a square increase uniformly at the rate of $\frac{1}{10}$ of a square inch per second, at what rate is the side increasing when the area is 100 square inches?

4. If the solidity of a cube increase uniformly at the rate of a cubic inch per second, at what rate is the length of the side increasing when the solid becomes a cubic foot?

(6.) To find a rule for integrating a differential of the form $x^{-\frac{1}{3}}dx$, $x^{\frac{2}{3}}dx$, $x^{\frac{1}{2}}dx$, &c. in which the index of x is fractional.

RULE. Add one to the index, divide by the index thus increased, and also by the differential of the variable.

$$\text{Ex.} \quad \int \tfrac{1}{2}x^{-\frac{1}{2}}dx = \frac{\frac{1}{2}x^{-\frac{1}{2}+1}dx}{\frac{1}{2}dx} = x^{\frac{1}{2}} + \text{C}.$$

EXERCISES.

1. What is the integral of $x^{-\frac{2}{3}}dx$?

2. What is the integral of $x^{\frac{1}{3}}dx$?

3. What is the integral of $\frac{dx}{\sqrt{x}}$?

4. What is the integral of $\frac{adx}{b\sqrt{x^{\frac{1}{3}}}}$?

5. The side of a square increases uniformly at the rate of $\frac{1}{10}$ of an inch per second; what is the area of the square when it is increasing at the rate of a square inch per second?

(7.) To find the differential of the sum or difference of two or more functions.

It is obvious that if the rates at which two quantities increase be added together, the sum will be the rate of increase at which the *sum* of the quantities increases; and the difference, the rate at which the *difference* of the quantities increases.

If there be two independent variables x and y, and if x increase whilst y decreases, we must consider the differential of y as *negative*.

Thus, the differential of $(ax + x)$ is $adx + dx$, and that of $(ax - x)$ is $adx - dx$.

EXERCISES.

1. What is the differential of $a^2x^2 + 2ax^3 - x$?
2. What is the differential of $(a + x)^2$?
3. What is the differential of $(a + x)(a - x)$?
4. What is the differential of $a\sqrt{x} + \frac{1}{2}y^2 - \frac{x}{3}$?

(8.) To integrate an expression consisting of any number of differentials connected by the signs *plus* or *minus*.

RULE. Take the integral of each separately, and connect them by their proper signs.

Ex. $\int (2xdx + x^2dx - dx) = x^2 + \frac{1}{3}x^3 - x.$

EXERCISES.

1. What is the integral of $(ax^2dx + \frac{dx}{2\sqrt{x}})$?

2. Integrate $(a^2x^{\frac{1}{3}}dx + \frac{dx}{x^{\frac{1}{4}}} - dy)$?

(9.) To differentiate a compound expression raised to the second or third powers.

Ex. To differentiate $(ax + x^2)^2$

Let $u = (ax + x^2)^2$, then $u^{\frac{1}{2}} = ax + x^2$,

and $\frac{1}{2}u^{-\frac{1}{2}}du = adx + 2xdx$. Therefore

$$du = \frac{adx + 2xdx}{\frac{1}{2}u^{-\frac{1}{2}}}, \text{ substituting the value of } u,$$

$$du = \frac{adx + 2xdx}{\frac{1}{2}(ax + x^2)^{1-2}} = 2(ax + x^2)^{2-1}(adx + 2xdx).$$

Rule. Multiply by the index, diminish the index by unity, and multiply by the differential of the compound expression.

Ex. Differentiate $(ax^2 + x^3)^3$,

$$d.(ax^2 + x^3)^3 = 3(ax^2 + x^3)^2(2axdx + 3x^2dx).$$

EXERCISES.

1. . What is the differential of $(1 + x)^2$?

2. What is the differential of $\left(\frac{x}{2} + \frac{ax^2}{b}\right)^3$?

3. What is the differential of $\left(\frac{1 + x}{a}\right)^2$?

4. If x increase uniformly at the rate of .01 per second, at what rate is the expression $(1 + x)^3$ increasing per second when x becomes 9?

(10.) To integrate an expression of the form $(ax + x^2)^2 (adx + 2xdx)$ in which the part following the compound expression inclosed in a parenthesis is the differential of the part inclosed.

Rule. Add one to the index of the compound expression, divide by the index thus increased, and also by the differential of the part inclosed in parentheses.

Ex. $\int (ax + x^2)^2 (adx + 2xdx) = \frac{1}{3} (ax + x^2)^3$

EXERCISES.

1. What is the integral of $(a + 3x^2)^2 6xdx$?
2. What is the integral of $(2x^3 - 1) 6x^2dx$?

(11.) To find the differential of the square or cube roots of a compound quantity.

Let $u = \sqrt{x + ax} = (x + ax)^{\frac{1}{2}}$,

Then $u^2 = x + ax$, and $2udu = dx + adx$.

Hence $du = \dfrac{dx + adx}{2u} = \dfrac{dx + adx}{2(x + ax)^{\frac{1}{2}}}$;

That is, $du = \frac{1}{2} (x + ax)^{\frac{1}{2}-1} (dx + adx)$.

Rule. Multiply by the fractional index, diminish the index by unity, and multiply by the differential of the compound expression.

EXERCISES.

1. What is the differential of $\sqrt{1 + x}$?
2. What is the differential of $\sqrt[3]{a + x^2}$?

3. Differentiate the function $\left(\frac{a}{b}+\frac{x}{2}\right)^{\frac{1}{3}}$.

4. Differentiate the function $\sqrt{x^2+a\sqrt{x}}$.

(12.) To find the integral of a differential of the form $(x + ax)^{\frac{1}{2}}(dx + adx)$, in which the differential part is the differential of the root of the compound expression.

Rule. Add one to the index, and divide by the index thus increased, and by the differential of the compound quantity.

Ex. $\int (x+ax)^{\frac{1}{2}}(dx+adx) = \frac{(x+ax)^{\frac{3}{2}}(dx+adx)}{\frac{3}{2}(dx+adx)}$

$= \frac{2}{3}(x+ax)^{\frac{3}{2}} + C.$

EXERCISES.

1. What is the integral of $\frac{dx}{(1+x)^{\frac{1}{2}}}$?

2. What is the integral of $\sqrt{a+x^2} \times 2xdx$?

SECTION IV

ON THE DIFFERENTIATION AND INTEGRATION OF EXPRESSIONS CONTAINING TWO INDEPENDENT VARIABLES.

It frequently happens in the solution of real problems that two variable quantities, whose rates of variation are perfectly independent of each other, enter into the investigation, and that the rate of variation of the function into which they enter is required to be determined. When these variable quantities, or the expressions into which only one of them enters, are connected by the signs *plus* or *minus*, the rate of variation of the compound function will be the sum or difference of the rates of its several terms. The only two cases then which we have to consider are those which involve the *products* or *quotients* of these variables.

(11.) To find the differential of the product of two variables xy.

Since $(x+y)^2 = x^2 + 2xy + y^2$, diff. both sides, $2(x+y)(dx+dy) = 2xdx + d.2xy + 2ydy$, or dividing by 2, and multiplying the terms of the first side together, $xdx + ydx + xdy + ydy = xdx + d.xy + ydy$, expunging the equal terms from both sides of the equation, we have $d.xy = xdy + ydx$.

Rule. Multiply each variable by the differential of the other, and add the products.

COR. To differentiate the product of any number of variables, multiply each differential by the product of the other variables, and take the sum of the several products.

Thus $d.\, xyz = xydz + xzdy + yzdx$.

For let $v = yz$, then $xyz = xv$, and $d.\, xv = xdv + vdx$, but $dv = ydz + zdy$. Hence, by substitution, $d.\, xyz = xydz + xzdy + yzdx$.

(2.) The rule may also be deduced from the following principle. If we consider one of the variables to retain its value whilst the other changes its magnitude, we obtain the rate at which the product varies. If we now supposed the last variable to remain constant whilst the other varies, we obtain the rate of variation of the product on this supposition. Hence, if both vary together, the sum of the partial rates of variation will give the actual rate of variation or the differential of the product. Thus, if x remain constant whilst y varies, the partial differential of the product will be xdy. Again, if y remain constant and x vary, the partial differential is ydx. Hence, when both vary $d.xy = xdy + ydx$.

EXERCISES.

1. If one side of a rectangle vary at the rate of 1 inch per second, and the other at the rate of 2 inches, at what rate is the area increasing when the first side becomes 8 inches and the last 12?

2. If one side of a rectangle *increase* at the rate of 2 inches per second, and the other *diminish* at the rate of 3 inches per second, at what rate is the area

increasing or diminishing, when the first side becomes 10 inches and the second 8?

3. What is the differential of xy^2?

4. What is the differential of $x^2y^{\frac{1}{2}}z^3$?

5. What is the differential of $ax^2y + x\sqrt{y} - \sqrt{xy}$?

(3.) If we have a differential of the form $xdy + ydx$, its integral will be xy.

Thus, $\int (xdy + ydx) = xy$.

EXERCISES.

1. What is the integral of $3x^2y^2dy + 2xy^3dx$?

2. What is the integral of $3xdy + 3ydx$?

(4.) To find the differential of the quotient of two independent variables, that is the differential of the fraction $\frac{x}{y}$.

Let $u = \frac{x}{y}$, then $uy = x$, and diff.

$$udy + ydu = dx, \text{ or } ydu = dx - udy.$$

Hence, $du = \frac{dx - udy}{y} = \frac{ydx - xdy}{y^2}$ by substitution.

Rule. From the differential of the numerator multiplied by the denominator, subtract the differential of the denominator multiplied by the numerator, and divide by the square of the denominator.

(5.) This rule will enable us to differentiate a variable quantity whether simple or compound, raised to a negative index.

Let it be required to differentiate x^{-2} or $\frac{1}{x^2}$

$$d.\frac{1}{x^2}=\frac{d.1\times x^2-2xdx\times 1}{x^4}=\frac{-2xdx}{x^4}=-2x^{-3}dx.$$

Hence, the rule is the same as formerly given for a positive index.

EXERCISES.

1. What is the differential of $\frac{x^2}{y^3}$?

2. What is the differential of $\frac{a+x}{a+x^4}$?

3. What is the differential of $\frac{xy}{a+x}$?

4. What is the differential of $\frac{1}{x}$?

5. What is the differential of $\frac{1}{\sqrt{x}}$?

6. What is the differential of $\frac{1}{\sqrt{a+x}^3}$?

(6) By observing the forms of the differentials of the above expression, the pupil will see by inspection, the particular fraction from which they resulted.

Thus, $\int \frac{ydx-xdy}{y^2}=\frac{x}{y}.$

EXERCISES.

1. Integrate $\frac{4nxdxy-2nx^2dy}{4y^2}$

2. What is the integral of $\frac{-dx}{a^2+2ax+x^2}$?

SECTION V.

ON THE DIFFERENTIATION AND INTEGRATION OF FUNCTIONS HAVING GENERAL INDICES.

In the preceding part we confined the attention of the pupil entirely to the simple forms of expression, in order that he might see more clearly the great leading principle of the calculus. We shall now show him that the rules obtained for the lower powers and roots, apply to all powers and roots of the same functions.

(1.) To find the differential of x^n.

Let $u=x^n$ and $u'=(x+h)^n$.

Then, $u'=x^n+nx^{n-1}h+n.\frac{n-1}{2}x^{n-2}h^2+$, &c. by the Binomial Theorem.

Hence, $u'-u=nx^{n-1}h+n.\frac{n-1}{2}x^{n-2}h^2+$, &c.

$$\frac{u'-u}{h}=nx^{n-1}+n.\frac{n-1}{2}x^{n-2}h+, \text{ \&c.}$$

Hence, taking the limit when $h=0$, we have $\frac{du}{dx}=nx^{n-1}$ and $du=nx^{n-1}dx$, which is exactly the same form as for x^2 or x^3.

(2.) To differentiate $x^{\frac{1}{n}}$

Let $u=x^{\frac{1}{n}}$, then $u^n=x$, raising both sides to the nth power. Hence $nu^{n-1}du=dx$, by differentiating, and $du=\frac{dx}{nu^{n-1}}=\frac{1}{n}x^{\frac{1}{n}-1}dx$.

(3.) To differentiate $\frac{1}{x^n}$, also $\frac{1}{x^{\frac{1}{n}}}$.

Since the differential of 1 is 0, if these expressions be differentiated by the rule for fractions, we have

$$d.x^{-n}= -nx^{-n-1}dx= -\frac{ndx}{x^{n+1}}$$

Also $d.x^{-\frac{1}{n}} = -\frac{1}{n}x^{-\frac{1}{n}-1}dx= -\frac{dx}{nx^{\frac{1}{n}+1}}$.

(4.) To differentiate $(a+x^2)^n$.

Let $u=(a+x^2)^n$, then $u^{\frac{1}{n}}=a+x^2$.

And $\frac{1}{n}u^{\frac{1}{n}-1}du = 2xdx$, and $du = \frac{2xdx}{\frac{1}{n}u^{\frac{1}{n}-1}}$.

Hence, $du=n(a+x^2)^{n-1}2xdx$, by substituting the value of u.

(5.) To differentiate $(a+x^2)^{\frac{1}{n}}$

Let $u=(a+x^2)^{\frac{1}{n}}$, then $u^n=a+x^2$,

and $nu^{n-1}du=2xdx$, $du=\frac{2xdx}{nu^{n-1}}$.

Hence $du=\frac{1}{n}(a+x^2)^{\frac{1}{n}-1}2xdx$.

(6.) To differentiate $\frac{1}{(a+x^2)^n}$ or $\frac{1}{(a+x)^{\frac{1}{n}}}$.

These expressions being differentiated by the Rule for fractions.

$$d.\frac{1}{(a+x^2)^n}, = -n(a+x^2)^{-n-1}2xdx = -\frac{2nxdx}{(a+x^2)^{n+1}}.$$

$$d.\frac{1}{(a+x^2)^{\frac{1}{n}}}, = -\frac{1}{n}(a+x^2)^{-\frac{1}{n}-1}2xdx = -\frac{2xdx}{n(a+x^2)^{\frac{1}{n}+1}}.$$

The rules for all these expressions may be thus stated.

Rule. Multiply by the index, diminish the index by unity, and multiply by the differential of the *root*.

EXERCISES.

1. What is the differential of ax^{n+1}?

2. What is the differential of $\frac{a}{b}x^{\frac{1}{n}}+C$?

3. What is the differential of $\frac{a}{x^n}$?

4. Differentiate $\frac{a}{bx^{\frac{1}{n}}} - C$?

5. Differentiate $(ax^n + x^m)^3$.

6. Differentiate $\frac{a}{(1-x)^n}$.

7. Differentiate $\frac{1}{a(a+x)^{\frac{1}{n}}}$.

8. Differentiate $\frac{x}{1-x}$.

9. What is the differential of $\frac{x}{yz}$?

10. What is the differential of $\frac{xy^n}{x^2-y^2}$?

Obs. The learner should write down the differentials which he obtains, shut the book, and endeavour by the *reverse* operation to obtain the integrals, which he will then compare with the above expressions to ascertain their identity.

SECTION VI.

ON THE REDUCTION OF DIFFERENTIALS TO KNOWN FORMS, INTEGRATION BY SERIES, AND DEFINITE INTEGRALS.

(1.) In the preceding part we have only considered the modes of finding the integrals of those differentials which were derived from functions of a particular form; the rules in all these cases being the *reverse* of those employed for differentiating. If the pupil meet with differentials which differ from the forms, with which he is now supposed to be familiar, *only* in being multiplied or divided by constant quantities, their differentials may easily be brought to the proper form without altering the value, and the integrals obtained. We shall illustrate these by examples.

Ex. 1. Required the integral of $(a+x^n)^2 x^{n-1} dx$?

In this example the part $x^{n-1}dx$ is not the differential of the root, or of $a+x^n$ which is $nx^{n-1}dx$. If we therefore multiply this part by n and divide by n, or place $\frac{1}{n}$ before the whole, the expression becomes

$$\frac{1}{n}(a+x^n)^2\, nx^{n-1}dx.$$

Hence,

$$\int \frac{1}{n}(a+x^n)^2 nx^{n-1}dx = \frac{1}{n}\int (a+x^n)^2\, nx^{n-1}dx$$

$=\frac{1}{3n}(a+x^n)^3$ the integral required.

The differential given in this example may be supposed to have been derived from differentiating $(a+x^n)^3$ and then dividing by n.

EXERCISES.

1. What is the integral of $(a+x^2)^3xdx$?
2. What is the integral of $(a^2+x^n)^3anx^{n-1}dx$?
3. What is the integral of $(ax+bx^2)^n(\frac{1}{2}ax+bnx)dx$?
4. What is the integral of $\frac{x^2dx}{(a+bx^3)^2}$?
5. Find the integral of $6yxdx+3x^2dy$.
6. Find the integral of $\frac{8yxdx+4x^2dy.}{y^2}$

(2.) When a differential cannot be brought under any of the preceding forms, we may sometimes be able to convert it, either immediately, or after certain substitutions, into an infinite series, which we may integrate by taking the integral of each term separately.

Ex. To find the integral of $\frac{dx}{a+x}$.

Since $\frac{dx}{a+x}=dx\times\frac{1}{a+x}$ we have

$$\frac{dx}{a+x}=\left(\frac{1}{a}-\frac{x}{a^2}+\frac{x^2}{a^3}-\frac{x^3}{a^4}+\text{\&c.}\right)dx,$$

or $\dfrac{dx}{a+x}=\dfrac{dx}{a}-\dfrac{xdx}{a^2}+\dfrac{x^2dx}{a^3}-\dfrac{x^3dx}{a^4}+$, &c.

Hence,

$$\int\frac{dx}{a+x}=\int\frac{dx}{a}-\int\frac{xdx}{a^2}+\int\frac{x^2dx}{a^3}-\int\frac{x^3dx}{a^4}+\text{ \&c.}$$

That is

$$\int\frac{dx}{a+x}=\frac{x}{a}-\frac{x^2}{2a^2}+\frac{x^3}{3a^3}-\frac{x^4}{4a^4}+, \text{ \&c.}$$

Hence, if we wish to find the value of the integral for any particular value of x, we take a sufficient number of terms, depending on the rapidity with which the series converges.

EXERCISES.

1. What is the integral of $\dfrac{dx}{1-x}$?

2. What is the integral of $\dfrac{dx}{1+x^2}$?

3. What is the integral of $\dfrac{dx}{\sqrt{1-x^2}}$?

(3.) When the *form* of an integral belonging to a given differential has been determined, the expression is called an *Indefinite Integral*, because we neither know the value of the independent variable nor that of the constant. But in the application of the calculus to the solution of *real* problems, the complete value of the integral is determined by the conditions of the problem. Since the value of the integral depends on that of the variable, we may determine the *value* of the constant, or make it disappear entirely from the integral by the following conditions. If we suppose the independent variable and the integral to *begin* to exist at the same instant, then when $x=0$, the integral $=0$ and consequently $C=0$.

Again, if we suppose the integral to begin to exist, or to have its *origin* when x becomes equal to a given quantity a, the value of C may then be determined.

Let the differential be $2nxdx$, then $\int nxdx=\frac{1}{2}nx^2+C$; let the origin of the integral be when $x=a$, then $\frac{1}{2}na^2+C=0$, and $C=-\frac{1}{2}na^2$, substituting the value of C, we have $\int nxdx=\frac{1}{2}nx^2-\frac{1}{2}na^2$. If we take another value of x, suppose $=b$, then the *definite* value of the integral will be $\frac{1}{2}nb^2-\frac{1}{2}na^2$.

When the value of the constant has been determined, and a *particular* value assigned to the independent variable, the value of the integral is then known, and is called a *Definite Integral.*

GEOMETRICAL ILLUSTRATION.

In fig. page 9, let the base AD of the triangle be called x, and let the perpendicular $DE=nx$, then the area of $ADE=\frac{1}{2}nx^2$, taking the differential we have $nxdx$. Let us now take the integral of $nxdx$, then $\int nxdx=\frac{1}{2}nx^2+C$. In this case the base x and the area begin to increase at the same time; hence, when $x=0$, $\frac{1}{2}nx^2+C=0$, and $C=0$.

Again, suppose we wish to ascertain an expression for the area of DD′E′E, contained between the two perpendiculars DE, D′E′; since the perpendicular is supposed to move uniformly along the line AD from the point A, where it is 0, the area of the figure DD′E′E does not *begin* to exist till the variable x becomes equal to AD; that is, when x=AD=a, the area or integral $\frac{1}{2}na^2+C=0$; hence $C=-\frac{1}{2}na^2$. Let AD′ be another value of x, which we may call b, then the integral of $nxdx$ or the area of AD′E′ will be expressed by $\frac{1}{2}nb^2+C$. Hence $\frac{1}{2}nb^2-\frac{1}{2}na^2=$ area of DD′E′E. The constant C may be made to disappear, or be eliminated, by giving two successive values to the independent variable and taking the difference between the two integrals corresponding to these values. Thus, if $x=a$ the integral is $\frac{1}{2}na^2+C$, if $x=b$ it is $\frac{1}{2}nb^2+C$, subtracting the former value from the latter we have $\frac{1}{2}nb^2-\frac{1}{2}na^2$ as before.

When we take the *excess* of the value of an integral when the independent variable has become equal to b, above its value when it was only equal to a, we are said to *integrate* between the limits of $x=a$, and $x=b$.

This is indicated by the sign $\int_a^b$.

Thus $\int_a^b nxdx=\frac{1}{2}nb^2-\frac{1}{2}na^2=\frac{b^2-a^2}{2n}$.

EXERCISES.

1. Integrate $\int_a^b 2xdx$, and illustrate the case by a geometrical example.

2. Integrate $\int_a^b 3x^2dx$, illustrate geometrically, and determine its numerical value when $a=4$ and $b=6$.

3. Integrate $\int_a^b \frac{\pi}{4} x dx$, illustrate by a geometrical example, and determine the definite integral when $a=2$ and $b=3$, π being 3, 1416.

4. Integrate $\int_a^b \frac{\pi}{2} x^2 dx$, illustrate geometrically, and determine the value of the definite integral between the limits $a=4$ and $b=6$.

OBSERVATIONS ON THE PRECEDING PART.

1. The learner must have observed, that when we attempt to ascertain the limits of certain fractional expressions, as they are given, we arrive at the unintelligible expression $\frac{0}{0}$. Fractions of this kind are called *Vanishing Fractions*. When he arrives at this expression he must not conclude that the fraction has no limit, but must proceed to determine its limit or ascertain that it has *really* no limit. If he can conveniently divide the numerator by the denominator, he may do so and take the limit of the quotient. He may sometimes arrive at the limit by substituting the rectangle contained by the sum and difference of two quantities, instead of the difference of their squares. Or, lastly, he may substitute $a+h$ for x, and find the limit when $h=0$.

Ex. Required, the limit of $\frac{x^4-a^4}{(x-a)}$, when $x=a$?

If we make $x=a$ without any preparation we have $\frac{0}{0}$, or the fraction is a vanishing fraction.

Dividing x^4-a^4 by $x-a$ we get $x^3+x^2a+xa^2+a^3$, which, when $x=a$, becomes $a^3+a^3+a^3+a^3$, or $4a^3$, which is the limit required. Or, let

$x=a+h$, then $\frac{(a+h)^4-a^4}{a+h-a} = \frac{4a^3h+6a^2h^2+4ah^3+h^4}{h}$

$=4a^3+6a^2h+4ah^2+h^3$. If we now make $h=0$, which is the same thing as making $a=x$, we have the limit $4a^3$ as before.

If after substituting $a+h$ for x, and going through the operation as above, we should arrive at an expression having h in all the terms of the numerator, but not in all the terms of the denominator, the numerator becomes 0, and the fraction has no limit, or we say its limit is 0. Again, if h should be found in every term of the denominator, but not in every term of the numerator, the denominator becomes 0, and the fraction has no limit, or its limit is said to be infinite, and is represented by the character ∞. Thus $\frac{a}{0}=\infty$.

There are other methods for finding the limits of such fractions, but they are not sufficiently elementary to be introduced in this place.

2. We cannot too strongly recommend to the teacher the necessity of giving the learner clear views of the nature of Differentials and the Differ-

ential Coefficient. The differential of the independent variable, when we are employing real quantities, is either a part of the variable itself, or we may take it equal to the whole variable, or even twice, three times, or any number of times of the variable. The differential of the function is also a part of the function, or it may be equal to it, or even greater.

Thus, if the side of a square increase uniformly at the rate of one inch per second, the area is increasing at the rate of $2xdx$, or $2x$, when the side becomes equal to x. If x be equal to 12 inches, then $2xdx =$ 24 square inches, which is the differential of the area.

Again, if the side increase uniformly at the rate of one foot per second, the area increases at the rate of $2 \times 12 \times 12 = 288$ square inches, which, on this supposition, is the differential of the area.

In the ratio $\frac{du}{dx} = 2x$, $2x$ is the differential-coefficient, which is equal to the differential of the function divided by the differential of the variable.

When $du = 24$ and $dx = 1$, $\frac{du}{dx} = 24$.

When $du = 288$ and $dx = 12$, $\frac{du}{dx} = \frac{288}{12} = 24$.

Hence, whatever be the rate of increase which we assign to the independent variable, the ratio of du to dx, or the differential coefficient, remains the same.

From all these illustrations the learner will perceive that the differential coefficient of a function, is the number which denotes the rate at which the function increases, the rate of increase of the independent variable being denoted by 1. Thus, if $u = x^2$, $\frac{du}{dx} = 2x = 18$, when $x = 9$; that is, the function x^2 increases 18 *times* faster than x when $x = 9$.

Again, if $u = \sqrt{x}$, $\frac{du}{dx} = \frac{1}{2\sqrt{x}} = \frac{1}{6}$, when $x = 9$. That is, the function $\sqrt{x}$ increases with $\frac{1}{6}$ of the rapidity that x increases when $x = 9$.

The differentials of quantities are therefore real quantities of the same kind with the variables themselves, and are never equal to 0, unless the quantities, of which they are the differentials, cease to be variable.

We have thus avoided calling on the learner to believe that the differential of u and the differential of x or du, dx, may each be equal to *nothing*, and nevertheless be to each other in a certain definite ratio. If we had tried to make him believe that $\frac{du}{dx} - \frac{0}{0} = 2x = \frac{2x}{1}$, he would very probably have put these quantities in the form of proportional numbers, after giving a particular value to x. Thus,

$$0 : 0 :: 2x : 1, \text{ or if } x = 10, \text{ as } 20 : 1.$$

If he happen to be a very intelligent boy, he will likely ask his teacher how *nothing* can be to *nothing* as 20 to 1, or how one of the *nothings* can be 20 times greater than the other?

As the learner will frequently meet with equations of this kind, $\frac{0}{0} = 2x = 3x^2 = nx^{n-1}$, &c. in almost every work on the differential calculus; and as he has yet been unable to attach any definite meaning to such expressions, he must now be told what mathematicians really mean by what appears to him an *absurdity*. If we take an improper fraction, as for example, $\frac{100}{10}$ which is equal to 10, and divide both numerator and denominator continually by 10, we rapidly diminish the numerator and denominator while the *value* of the fraction remains unchanged.

Thus, $\frac{.00000000001}{.000000000001} = 10$, also

$$\frac{.0000000001}{.000000000000001} = 100000.$$

That is, the numerator and denominator may be made as small as we please, whilst the value of the fraction may be as great as we please. By considering 0 as the *limit* at which numbers cease to exist, we say that we can make the numerator and denominator approach as near to 0 as we please, without altering their ratio, or the value of the fraction. Hence, by an extension of the meaning of the terms *limit* and *equality*, for which the learner is in general not prepared, the expression is generalised, and the equation

$\frac{0}{0} = 2x = 3x^2 =$ *nothing* $=$ *anything*, is obtained.

The pupil will now clearly perceive that what he believed to be an equation, is not an equation in the ordinary meaning of the term; but an abbreviated mode of expressing a truth with which he was perfectly familiar when studying vulgar and decimal fractions; namely, "That the value of a fraction does not depend on the *absolute* values of the numerator and denominator, nor on their *difference*, but merely on the number of times or '*parts of a time*' which the one is contained in the other."

EXERCISES.

1. What is the limit or value of the fraction $\frac{ax}{x}$ when $x=0$?

2. What is the limit of the fraction $\frac{ax^2}{x}$ when $x=0$?

3. What is the limit of the fraction $\frac{ax}{x^2}$ when $x=0$?

PROMISCUOUS QUESTIONS AND EXERCISES.

1. What do you understand by variable and constant quantities?
2. Have all variable quantities certain limits within which they may increase or diminish?
3. State what you understand by a Limit, and give a simple example.
4. What is meant by the Function of a variable quantity?
5. What common word or form of words may you employ instead of this technical term?
6. Why is one of the variable quantities called the independent variable and the function the dependent variable?

7. Is the independent variable supposed to increase or diminish at a uniform rate?

8. Does the function into which it enters increase or decrease uniformly when the independent variable does so?

9. If the function contain any power or root of the variable, does it increase uniformly?

10. What do you understand by the fluxion or differential of the independent variable?

11. If the independent variable be supposed to be divided into any number of equal parts, may you take one of those parts to represent its differential?

12. What do you understand by the differential of a function which does not increase uniformly?

13. What do you understand by the differential coefficient of a function?

14. What is the general rule for finding the coefficient in all the algebraic expressions you have employed?

15. State in your own words what is the object of the Differential Calculus.

16. What do you understand by the term Integral?

17. What is the object of the Integral Calculus?

18. Whether are the Rules you have yet learned for integration obtained from a general principle or an inverse operation?

19. What do you understand by the term Increment?

20. Are the differentials of quantities so exceedingly small, that they are about to vanish, or may they be of any magnitude we please?

21. If an independent variable be only an inch in length, may its differential be equal to a foot?

22. Does the differential coefficient of a function depend on the *absolute* magnitude of the differential of the variable?

23. When we speak of a variable quantity increasing or diminishing, is the *idea* of *motion* necessarily involved?

24. When we speak of a function *generally*, do all the letters and their powers, or roots represent abstract numbers, or numbers of different kinds?

25. If the side of an equilateral triangle increase uniformly at the rate of 3 feet per second, at what rate is the area increasing when the side becomes 10 feet?

26. If the side of an equilateral triangle increase uniformly at the rate of .001 of an inch per second, at what rate is its perpendicular increasing when the side is just passing the limit of 10 inches?

27. If a halfpenny be placed on a hot shovel, so as to expand uniformly, at what rate is its *surface* increasing when the diameter is passing the limit of 1 inch and $\frac{1}{10}$, the diameter being supposed to increase *uniformly* at the rate of .01 of an inch per second?

28. If a circular plate of metal expand by heat so that its *area* increases uniformly at the rate of .001 of a square inch per second, at what rate does its diameter increase when the area of the circle is exactly a square inch?

29. If the diameter of the plate expand uniformly at the rate of .001 of an inch per second, what is the *diameter* of the circle when its *area* is expanding at the rate of a square inch per second?

30. A boy with a mathematical turn of mind observing an idle boy blowing small balloons with soap-suds, asked him the following pertinent question:—If the diameter of these balloons increase uniformly at the rate of $\frac{1}{10}$ of an inch per second, at what rate is the internal capacity increasing at the moment the diameter becomes 1 inch?

31. The young experimental philosopher, not being so ignorant as the other supposed, determined the rate correctly, and put the following question to the other:—If I blow equal quantities of air into the ball in equal times, at what rate is the diameter increasing at the moment it becomes an inch, on the supposition that I am blowing air into the ball at the rate of a cubic inch per second?

32. As one problem frequently gives rise to another, the pupil, after *differentiating* and calculating, succeeded in answering the question, and then put the following:—If the diameter increase uniformly, what is the size of the ball when the rate of increase of the diameter is to the rate of increase of the capa city as 1 to 50?

33. If an ingot of silver of the form of a parallelopiped, expand uniformly by heat in its linear dimensions at the rate of .001 of an inch per second, at what rate is the solidity increasing when the breadth is 4 inches, the depth 3, and the length 12?

34. A boy standing on the top of a tower, whose height is 60 feet, observed another boy running towards the foot of the tower at the rate of 6 miles an hour on the horizontal plane: at what rate is he approaching the first when he is 100 feet from the foot of the tower?

PART II.

APPLICATIONS OF THE PRECEDING RULES AND PRINCIPLES TO USEFUL PURPOSES.

SECTION I.

ON THE MAXIMA AND MINIMA OF QUANTITIES.

(1.) If a variable quantity gradually increase, and after it has reached a certain magnitude gradually decrease, its greatest value is called a maximum. Thus in the annexed figure, if a line move from A along AB so as to be always at right angles to AB, whilst its extremity moves along the arc of the circle, it will at first increase very rapidly and gradually slower and slower till it reach the position CD, when it becomes equal to the radius of the circle. By continuing the motion, it will from that position decrease till it reach the point B, when it be-

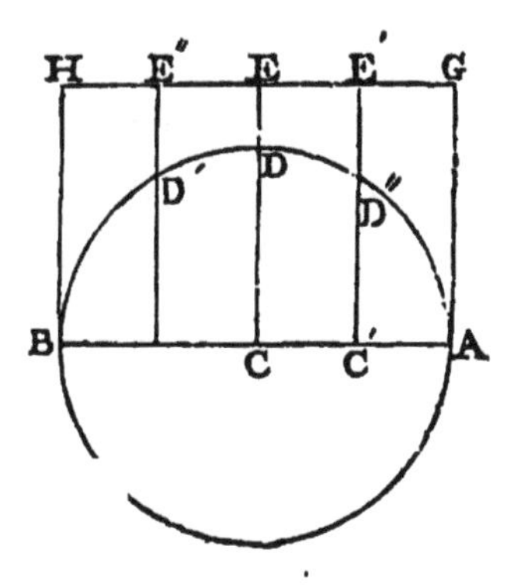

comes nothing, or vanishes. Hence, the line is said to be a maximum, or at its *greatest* value, when it becomes equal to the radius of the circle.

Again, if we consider the line AG moving along GH in the same manner, its extremity moving in the arc of the circle, it will continually diminish till it reach the position ED, after which it will increase till it arrive at BH. Hence, in the position ED it is said to be a minimum, or at its *least* possible value.

(2.) Now as the differential of a quantity is the rate of its increase or decrease, it is obvious that when a quantity reaches its greatest or least magnitudes, it neither *increases* nor *diminishes*, and, consequently, its differential is 0.

(3.) When a quantity becomes a maximum or minimum, it is obvious that its square, cube, square root, &c. will be greater or less than in any other state, and will therefore be a maximum or minimum. It is equally obvious that any number of times, or any part of the variable in its greatest or least states, will be the greatest or least values of the variable.

(4.) In the solution of problems, or other investigations involving the consideration of variable quantities in their greatest or least magnitudes, we must obtain an algebraic expression for the quantity, take its differential and form an equation, having the differential on one side and 0 on the other, from which the value of the independent variable x will be obtained. A few examples will illustrate these remarks, and delight the pupil with his newly acquired powers of calculation.

EXERCISES.

1. Divide a line into two parts so that the rectangle contained by the parts may be a maximum. Let a be the line, x one of the parts, and $a-x$ the other.

Then $(a-x) = ax-x^2 =$ a max.

Differentiating $adx-2xdx = 0$.

$$adx = 2xda.$$

Hence, $x = \frac{1}{2}a$, that is, the line must be bisected.

COR. Hence, of all rectangles having the same perimeter, the square has the greatest area.

2. Divide a given line, AB, into two parts, so that the sum of the areas of the squares described on these parts shall be the least possible.

Let $a =$ the line x, one of the parts, then $a-x$ will be the other part. Then,

$$x^2+(a-x)^2 \text{ will be a minimum.}$$

That is, $2x^2+a^2-2ax$ is a min. hence

$4xdx-2adx=0$, and $4xdx=2adx$,

$$\text{And } 4x=2a, \text{ and } x=\tfrac{1}{2}a.$$

Hence the line must be bisected.

3. A gentleman has a park in the form of a triangle, the base of which is 400 feet and the per-

pendicular 300, in which he wishes to make a rectangular garden, one of the sides of which is to be on the base; it is required to find how many feet from the vertex the other side must be drawn?

As the numbers 400 and 300 are constant, we may represent them by a and b. Let CK $= x$, then DK$=b-x$, then by the property of similar triangles we have $b : x :: a :$ GF. Hence GF$=\frac{ax}{b}$, consequently the area of the rectangle is equal to

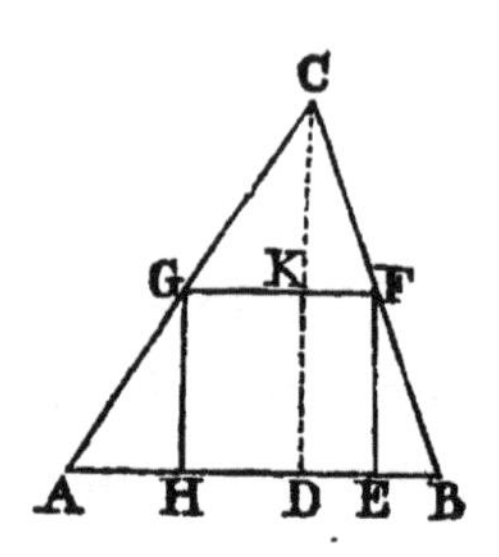

$$\text{GF} \times \text{DK} = \frac{ax}{b} \times \overline{b-x} = \frac{a}{b}(bx-x^2) = \text{a maximum.}$$

But since $\frac{a}{b}$ is constant, the quantity $bx-x^2$ will also be a maximum. Hence $bdx-2xdx = 0$, and $x = \frac{b}{2}$. Hence the perpendicular must be bisected.

4. Let AB be the diameter of a given circle, it is required to find a point, C, in the diameter, so that the rectangle formed by the chord DE, which is perpendicular to AB, and the part AC may be the greatest possible.

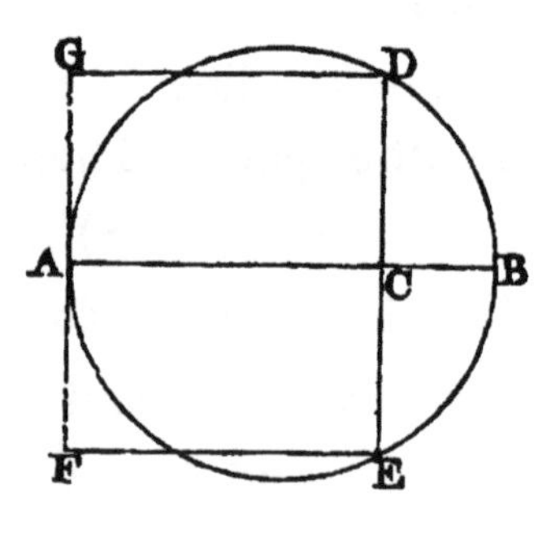

Let AB $= a$, AC$=x$, and CB $= a-x$, then

$$(a-x)x = \text{CD}^2 \text{ and CD} = \sqrt{ax-x^2};$$

therefore DE $= 2\sqrt{ax+x^2}$, and the rectangle EG $=$ $x \times 2\sqrt{ax-x^2}$, which is a maximum. Hence, since $2x\sqrt{ax-x^2}$ is a maximum, its square will also be a maximum. That is, $4x^2(ax+x^2)$, or $4ax^3-4x^4$ is a maximum. Differentiating this expression, we have

$$12\,ax^2dx - 16x^3dx = 0\,;$$

That is, $12\,ax^2dx = 16x^3dx$.

$$\therefore\ x = \frac{3}{4}a.$$

5. A tin-smith was ordered to make a cylindrical vessel, which should hold a given quantity, a gallon for example, and be the *lightest* possible; required the proportion between the radius of its base and height?

Since the tin-plate is of uniform thickness, the problem reduces itself to the finding the dimensions of the cylindrical vessel with the least possible surface.

Let c denote the capacity of the cylinder in cubic inches, x the radius of the base, and y the height of the cylinder, and $\pi = 3.1416$.

Then $\pi x^2 =$ area of base, $2\pi xy =$ area of the convex surface, and $\pi x^2 y = c$.

Hence, dividing both sides of this eq. by x we get

$$\pi xy = \frac{c}{x} \text{ or } 2\pi xy - \frac{2c}{x},$$

another expression for the convex surface.

Hence, $\pi x^2 + \frac{2c}{x} =$ whole surface, which is a minimum. no surf^ce at top. vessel is open.

Differentiating we have

$$2\pi xdx - \frac{2cdx}{x^2} = 0$$

Hence, $2\pi x dx = \frac{2cdx}{x^2}$ and $2\pi x^3 = 2c$,

Consequently, $x = \sqrt[3]{\frac{c}{\pi}}$

Again, since $\pi x^2 y = c$ we have $y -= \frac{c}{\pi x^2}$.

Therefore,

$$y = c \div \frac{\pi c^{\frac{2}{3}}}{\pi^{\frac{2}{3}}} = \frac{c^{\frac{3}{3}}}{\frac{\pi c^{\frac{2}{3}}}{\pi^{\frac{2}{3}}}} = \frac{{}^{\frac{1}{3}}}{\pi^{\frac{1}{3}}} = \sqrt[3]{\frac{c}{\pi}}$$

Consequently, by substituting the value of x we have

$$y = \frac{c}{\pi \frac{c^{\frac{2}{3}}}{\pi^{\frac{2}{3}}}} = -\frac{c^{\frac{3}{3}}}{\pi \frac{c^{\frac{2}{3}}}{\pi^{\frac{2}{3}}}} = \frac{c^{\frac{1}{3}}}{\pi^{\frac{1}{3}}} = \sqrt[3]{\frac{c}{\pi}}$$

which being the same expression as that obtained for x, it follows the height of the cylinder must be equal to the radius of its base.

6. In a very ungraceful spire of a church, the last stone was a large cone, whose base was 4 feet and height 10; the spire being taken down, and this stone being of no use in its present shape, it occurred to the clergyman that it might be converted into an excellent roller for smoothing the walks in his garden; required the length of the cylinder so as to have the greatest weight possible.

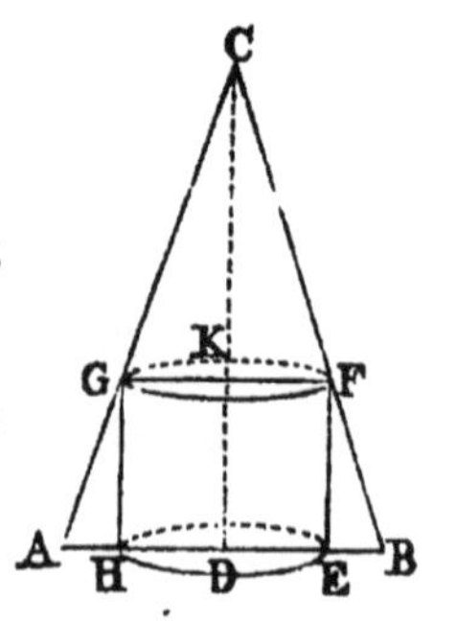

Let the cylinder be supposed cut by a plane passing through the vertex and also through the centre, and let CD $= a$, AB $= b$, CK $= x$, then KD $= a - x$.

By similar triangles, $a : b :: x :$ FG,

hence FG $= \frac{bx}{a}$

Again, since the solidity of a cylinder is equal to the area of its base multiplied by its height, the solidity will be expressed by $HE^2 \times .7854 \times DK$, or using $\frac{\pi}{4}$ for .7854, we have $\frac{\pi}{4} \frac{b^2x^2}{a^2}(a-x) =$ a maximum;

or neglecting the constant multiplier $\frac{\pi}{4} \times \frac{b^2}{a}$

we have $ax^2 - x^3 -$ a maximum.

Hence, $2\,axdx - 3x^2dx = 0$.

Consequently, $2axdx = 3x^2dx$.

And $x = \frac{2}{3}a$, therefore $DK = \frac{1}{3}a$.

EXERCISES.

1. Divide 12 into two parts, so that the least multiplied by the square of the greatest, shall be a maximum.

2. Divide 12 into two parts, so that the greatest multiplied by the *cube* of the least, shall be a maximum.

3. The length of the hypothenuse of a right-angled triangle is 120 feet; required the length of the base and perpendicular, so that the area may be a maximum.

4. A person wishes to enclose a field, which shall contain a given area a, and be in the form of a right-angled triangle; required the base and perpendicular, so that their sum may be the *least* possible.

5. A gentleman having a circular fish-pond, whose diameter is 100 feet, wishes to enclose it by a fence in the form of an isosceles-triangle, the sides of which shall be tangents to the circles; required the length of the sides so as to enclose the *least* possible land along with the pond.

6. If the side of a square pyramid be 40 feet, it is required to find the height to which it must be built, so as to contain the *greatest* quantity of masonry and expose the *least* possible surface to the action of the air.

SECTION II.

ON CURVES OF THE SECOND ORDER.

Before entering on the application of the Calculus to Curves, &c. the pupil ought to have some knowledge of the Conic Sections; but for the advantage of those who may not have time to study large treatises, we shall exhibit the nature and construction, or *genesis* of these curves, with their most essential properties, called their *Equations*.

(1.) A curve is a continuous line which gradually changes its direction, so that no part of it, however small, is a straight line. The curve is thus said to observe the *law of continuity*, which all variable quantities are supposed to observe. By the expression *law of continuity* we mean, that there is no abrupt start from one magnitude to another, nor from one direction to another. If the same quantity be found to have two different magnitudes in succession, it is supposed to have passed *gradually*, though perhaps not uniformly, through *all* the intermediate magnitudes. Again, if a point be found in two successive instants of time to be moving in different directions, it is supposed to have passed gradually through all the intermediate directions.

(2.) If a curve have two equal branches, the straight line which lies symmetrically between them is called the *Axis* of the curve, and the point where it crosses the axis, the *Vertex.* If from any point in the curve a perpendicular be drawn to the axis, it is called an *Ordinate*, and its distance from the vertex, the *Abscissa.* An abscissa and its ordinate are called the *Co-ordinates* of the point in the curve. The abscissa is always denoted by x, and the ordinate by y.

(3.) An equation containing the ordinate and abscissa, combined with known quantities, is called the *Equation* of the curve. Thus, if a be the diameter of a circle, $y^2 = (a-x)x$ is the equation of the circle.

(4.) If the equation of a curve be given, the curve may be described by ascertaining a sufficient number of points which must all lie in the curve, and then tracing the curve through these points. Thus, if we take $x=1, 2, 3,$ &c. and find the corresponding values of y from the preceding equation, and draw a series of ordinates at these points of the proper length, the curve which may be drawn through these points will be a semicircle.

(5.) The circle, from the simplicity of its construction and the uniformity of the curve, and the necessity of an early knowledge of its properties, is always placed in the elements of Geometry. The curves next in simplicity of their construction and equations are the *conic sections*. They are often called lines of the *second order*, because the highest power of any quantity in their equations is the *square* or *second* power.

I. THE PARABOLA.

1. Let AB be a straight line and CD another, at right angles to it, and let a point E move in such a manner that its distance from a given point F, and its perpendicular distance from the line AB, shall always be equal, the point will describe a curve called a *parabola*. The line AB is called the *directrix*, the point F the *focus*, V the *vertex*. The double ordinate EG passing through the focus is called the *parameter*, or *latus rectum*.

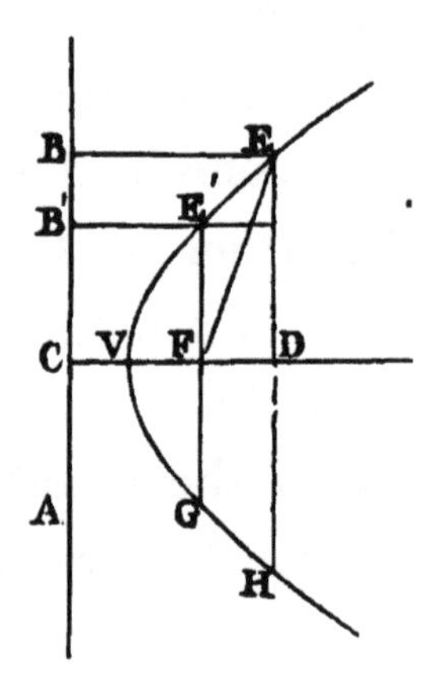

(2.) To find the equation to the curve.

From the genesis of the curve $CV = VF$ and $FE' = E'B'$ and $E'G = 2CF = 4CV =$ parameter p.

Draw the ordinate ED, put $VD = x$ and $DE = y$.

Then $DE^2 = FE^2 - FD^2 = BE^2 - FD^2 = CD^2 - FD^2 =$

$$= (CD + FD)\ (CD - FD) = 2VD \times CF = VD \times 2CF$$

That is $y^2 = px$, the equation required.

Cor. Let y' be any other ordinate and x' its abscissa.

Then $x : x' :: y^2 : y'^2$. Why?

Obs. Every curve whose abscissæ are proportioned to any power of the ordinates is called a parabola.

If $x : x' :: \sqrt{y^3} : \sqrt{y'^3}$, or $:: y^{\frac{3}{2}} : y'^{\frac{3}{2}}$, the curve is called the *semicubical parabola.*

The general equation of the parabola is $ax = y^n$.

DESCRIPTION, OR GENESIS OF THE CURVE.

1. Draw any number of lines parallel to AB, from F with the distance FC cut the parallel EG, in the points E′, G, and these will be two points in the curve. From F, with a radius equal to DC, describe arcs cutting EH in the points E, H, &c. Bisect CF in V and through the points V, E′, E, &c. draw a curve, which will be a parabola.

2. Join together two flat pieces of wood EC, CD, so as to form a right angle. Fix a thread at D, make it equal to DC, and form a small loop at the end C. Fix a pin in the paper at F, passing through the loop, then move the square

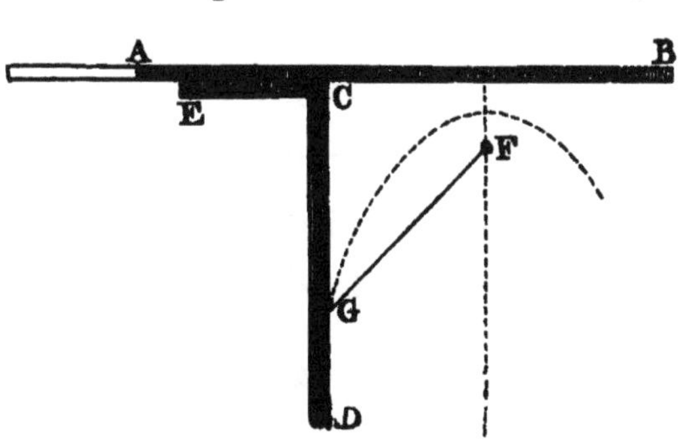

ECD along the edge of the slip of wood AB, keeping the thread applied to the edge of the square from D, by the point of a pencil G, and the point will trace the parabola by continued motion. For in every position the thread FG is equal to the part of the square GC to which it was applied.

II. THE ELLIPSE.

1. If a point G move in such a manner that the sum of its distances from the two given points F, f, shall always be the same, it will describe a curve called the ellipse. Each of the points F, f is called a *focus*. AB the *major axis* or *transverse diameter*; DE the *minor axis* or *conjugate diameter*. CF is called the *eccentricity*.

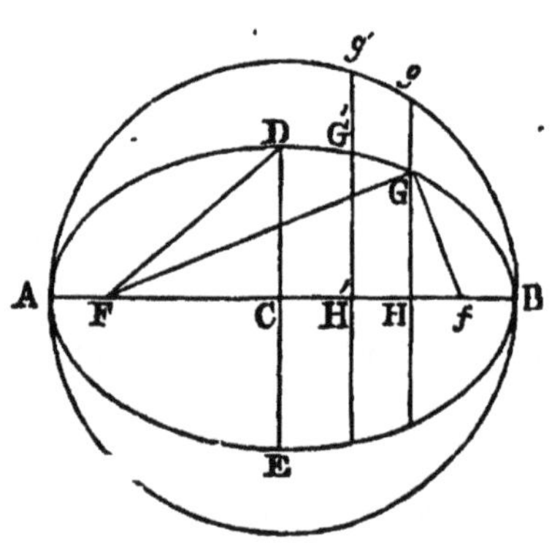

From the nature of the curve, AB=FG+fG=2FD.

2. To find the equation to the curve.

Let fall the perpendicular GH, and let AC$=a$, CD$=b$, CF$=e$, CH$=x$, HG$=y$ and let FG$-$FD$=d$. Then Ff $=2e$, FH$=e+x$ and fH$=e-x$. Also by the nature of the curve FG$=a+d$ and fG$=a-d$. Since FHG, fHG are right-angled triangles, we have

$$\text{FG}^2=\text{FH}^2+\text{HG}^2 \text{ and } f\text{G}^2=f\text{H}^2+\text{HG}^2.$$

$$\text{or } a^2+2ad+d^2+y^2=e^2+2ex+x^2+y^2 \ldots\ldots(1)$$

$$\text{and } a^2-2ad+d^2+y^2=e^2-2ex+x^2+y^2 \ldots\ldots(2)$$

By subtraction, $4ad=4ex$, and $d=\frac{4ex}{4a}=\frac{ex}{a}$.

Substituting this value in equation, (1) we have

$a^2+2a\times\dfrac{ex}{a}+\dfrac{e^2x^2}{a^2}+y^2=e^2+2ex+x^2+y^2$, by reducing, and substituting a^2-b^2 for e^2.

$$a^2y^2=b^2(a^2-x^2),\ \text{and}\ y^2=\frac{b^2}{a^2}(a^2-x^2)$$

$$\text{or, } y^2-\frac{b^2}{a^2}(a+x)(a-x)=\frac{b^2}{a^2}\times \text{AH}\times\text{BH},$$

If AH be denoted by x' then BH $=2a-x'$, and the equation becomes $y^2=\dfrac{b^2}{a^2}(2ax'-x'^2)$, which is the equation required.

Cor. 1. If the equation be converted into a proportion we have

$$\text{AC}^2 : \text{CD}^2 :: \text{AH}\times\text{HB} : \text{HG}^2.$$

Cor. 2. If a circle be described on AB and HG produced to g', and any other ordinate H$'g'$ drawn, then

since AH $\times$ HB $=$ Hg^2 and AH$'\times$ H$'$B $=$ H$'g'^2$.

we have HG^2 : Hg^2 :: H$'$G$'^2$: H$'g'^2$.

or HG : Hg :: H$'$G$'$: H$'g'$.

Cor. 3. Also HG : Gg :: H$'$G$'$: G$'g'$. Why?

GENESIS OF THE CURVE.

Take a thread equal to AB having two small loops at the extremities. Fix two pins perpendicularly in the paper at F,f, and passing through the loops, then with the point of a pencil trace the curve, keeping the thread FGf, always extended.

III. THE HYPERBOLA.

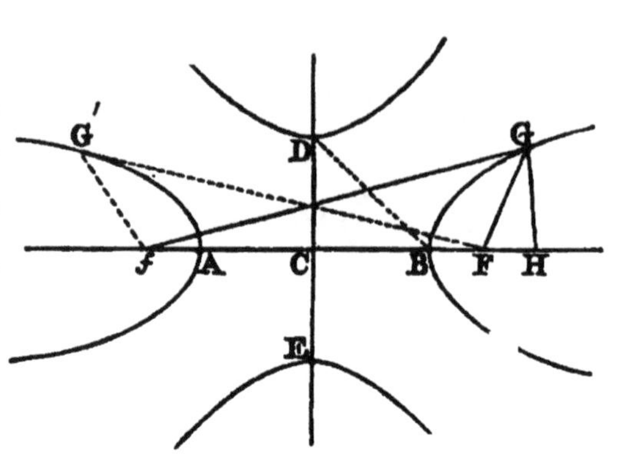

1. If a point, G, move in such a manner that the *difference* of its distances, fG, FG, from two given points, f, F, shall be always the same, the point will describe a curve, called the *Hyperbola*. If the point G′ move in the same manner, so that FG′ $-$ fG′ $=$ fG $-$ FG, it will describe a similar curve. These are called *Conjugate Hyperbolas.* The points F, f are called the *Foci*, AB the *Transverse Axis*, and the middle of AB or C the *Centre.* If DE be drawn at right angles to AB, and if from B, with a radius equal to CF, an arc be described cutting the line in D, E, then D E is called the *Conjugate Axis.*

From the nature of the curve fG $-$ FG $=$ fB $-$ FB $=$ AB.

2. To find the equation of the curve.

Let AB $=a$, ED $=b$, CF $=$ Cf $=e$, CH $=x$, HG $=y$, FG $=z$, then FH $=x-e$, fF $=2e$. And by the nature of the curve fG $=a+z$.

Now CB^2 + CD^2 = BD^2, or $\frac{1}{4}a^2+\frac{1}{4}b^2=e^2$; ... (1)

Also fG^2 = FG^2 + fF^2 + 2fF × FH, or substituting the single letters,

$$a^2+2az+z^2=z^2+4e^2+4ex-4e^2;\text{ hence } z=\frac{4ex-a^2}{2a},$$

$$\frac{16e^2x^2-8ea^2x+a^4}{4a^2}=z^2=\text{HG}^2+\text{FH}^2=y^2+x^2-2ex+e^2.$$

Hence, by reducing and simplifying, $16e^2x^2+a^4-4a^2y^2+4a^2x^2+4a^2e^2$, substituting the values of $16e^2$ and $4e^2$, which we obtain from equation, (1), and simplifying the results, we have

$$a^2y^2=b^2x^2-\tfrac{1}{4}a^2b^2=b^2(x^2-\tfrac{1}{4}a^2)=b^2(x+\tfrac{1}{2}a)(x-\tfrac{1}{2}a).$$

Let BH be denoted by x', the equation becomes $a^2y^2=b^2(a+x')x'$, or $y^2=\frac{b^2}{a^2}(a+x')x'$, which is the equation required.

COR. $\text{CB}^2 : \text{CD}^2 :: \text{AH}\times\text{BH} : \text{HG}^2$.

GENESIS OF THE CURVE.

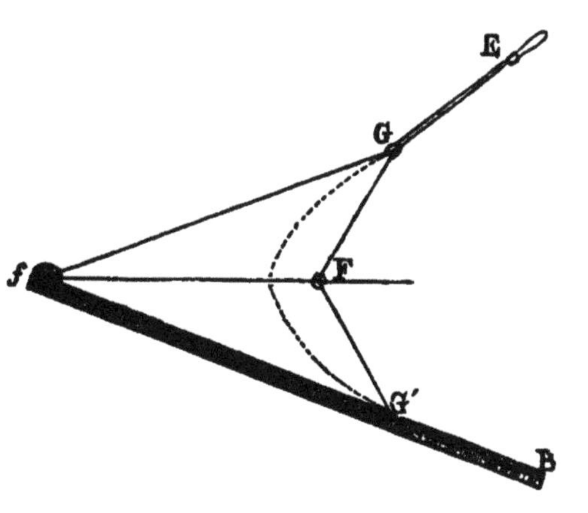

1. Take a thread with two small loops at the extremities; double it so as to leave one of the parts longer than the other, by a part which is to be the transverse axis, and tie a knot on the double part at E. If pins be fixed at F, f, passing through the loops, and the double thread be passed through a bead, or through a hole near the point of a pencil, and if the bead G or pencil be moved, keeping the thread tight, it will obviously describe the hyperbola.

2. The curve may also be described by the following method. Take a small slip of wood, fB; make a small

hole at *f*, and push a pin through it into the paper. Fix one end of a thread at B, the other end having a loop at F, through which a pin is to be fixed in the paper. The thread BG′F ought to be less than *f*B by a part equal to the transverse axis. If a pencil be now applied at G′, keeping the thread in contact with the edge of the slip of wood, and if the rule be moved round *f*, the point G′ will trace out the hyperbola.

OBS. The learner cannot fail to observe the striking resemblance between the equations of the circle, ellipse and hyperbola, the equation for the circle being $y^2 = ax-x^2$, for the ellipse $y^2 = \frac{b^2}{a^2}(ax^2-x^2)$; for the hyperbola, $y^2 = \frac{b^2}{a^2}(ax^2+x^2)$. If the foci of the ellipse approach one another, the transverse and conjugate axes approach to a ratio of equality, and $\frac{b^2}{a^2}$ also approaches to a ratio of equality, and when they become equal the curve becomes a circle. Hence the circle is, on this supposition, the *limit* to the ellipse.

In the ellipse the transverse axis is equal to the *sum* of the focal distances of any point in the curve; in the hyperbola it is equal to the *difference* of those distances. In the ellipse the conjugate axis b is equal to $\sqrt{\frac{a^2}{4}+e^2}$, e being the eccentricity. In the hyperbola it is equal to $\sqrt{e^2-\frac{a^2}{4}}$, the only difference being in the signs. Hence the properties of the ellipse may be transformed into the corresponding properties of the hyperbola by the proper change of plus into minus, or minus into plus.

Thus, the equation of the ellipse is $\frac{b^2}{a^2}(ax-x^2)$ or

$\frac{b^2}{a}(a-x)x$; now, in the hyperbola, if x be the abscissa, $a+x$ is the other abscissa, or that reckoned from the other extremity of the transverse, which substituted for $a-x$ in the equation of the ellipse, gives the equation of the hyperbola.

We have shown that the circle is the *limit* to the ellipse when the foci approach and ultimately coincide; what is its limit when their distance increases, and ultimately becomes infinite? In this case x is indefinitely small compared with a, and a and b approach to a ratio of equality. Hence y^2 approaches to the value of ax, which is its *limit*, and which is the equation to the parabola. Hence the parabola is the *limit* of the ellipse. When the transverse axis is very great, as in the case of the orbits of the comets, a large portion of the elliptic orbit nearest the sun almost coincides with a parabola; and hence, when near its perihelion, a comet may be considered as moving in a parabolic orbit.

SECTION III.

TANGENTS TO CURVES.

(1.) To determine the method of drawing tangents to curves.

Let AB be the axis of the curve and EF the tangent at E, meeting BA produced in F. Draw the ordinate ED and other two hK, h'K$'$ at equal distances from it.

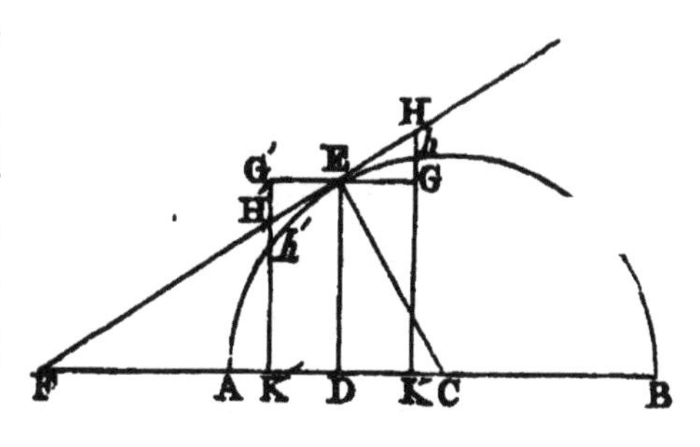

Draw EG at right angles to HK and EG′ to Hh' produced. Let an ordinate be supposed to move uniformly along AB, so that any part of AB as DK′ or DK may be taken to represent the rate at which the abscissa AD increases. Had the ordinate increased uniformly, GH would have been its corresponding increment, whereas it is only Gh. If the line KH be now supposed to move uniformly towards ED, the ratio of Gh to GH continually approaches to a ratio of equality. For since Eh is a curve, Gh diminishes more *slowly* than GH. Hence, these lines continually approach to a ratio of equality. When the ordinate has passed ED, its increments diminish more *rapidly* than if its extremity had kept in the tangent. Hence, when the ordinate was just passing ED, the rate of its diminution was to the rate of diminution of the abscissa as HG to EG, or as ED to DF. Hence, if DK or EG represent the differential of the abscissa, GH will represent the differential of the ordinate. Let AD$=x$, DE$=y$, then DK$=dx$ and GH$=dy$.

Hence, by similar triangles $dy : dx :: y :$ DF.

Consequently, DF $= \frac{ydx}{dy}$, which is the expression for the subtangent DF.

In applying this method to any particular curve, we obtain the expression $\frac{ydx}{dy}$, (from differentiating the equation to the curve,) equal to a known quantity, which determines the point F. Hence, if F,E be joined the line FE will be the tangent required.

Ex. To draw a tangent to a circle.

Let r be the radius, x the abscissa, and y the ordinate.

Then $y^2 = (2r-x)x = 2rx-x^2$ and $y = \sqrt{2rx-x^2}$

Differentiating this equation,

$$dy = \tfrac{1}{2}\,(2rx-x^2)^{-\frac{1}{2}}(2r-2x)dx.$$

Hence, $$\frac{dx}{dy} = \frac{1}{\tfrac{1}{2}(2rx-x^2)^{-\frac{1}{2}}(2r-2x)} = \frac{(2rx-x^2)^{\frac{1}{2}}}{r-x}$$

Multiplying the first side of the equation by y, and the second by its equal $\sqrt{2rx-x^2}$ we have

$\dfrac{ydx}{dy} = \dfrac{2rx-x^2}{r-x}$, the subtangent required.

EXERCISES.

1. If the diameter of a circle be 10 feet and the abscissa AD$=2$; required the length of the subtangent DF, and also of the tangent FE.

2. If a be half the transverse axis and b half the conjugate axis of an ellipse; required the length of the subtangent, when $x = 10$, a being $= 15$ and $b = 8$.

3. Draw a tangent to the hyperbola.

4. If the parameter of a parabola be 4 inches and the abscissa 9; required the length of the ordinate and subtangent.

SECTION IV.

NORMALS AND SUBNORMALS.

(1.) To find the length of the normal and subnormal of any point in a curve.

If a line EC (fig. p. 65) drawn at right angles to the tangent, meet the axis in C, the line EC is called the *normal*, and DC the *subnormal*.

Since the triangles EGH, EDC are similar,

EG : GH :: ED : DC; that is,

$$dx : dy :: y : \frac{ydy}{dx} = \text{DC, the subnormal.}$$

Again, since

$$\text{CE}^2 = \text{ED}^2 + \text{DC}^2 = y^2 + \frac{y^2 dy^2}{dx^2} = y^2 \left(1 + \frac{dy^2}{dx^2}\right)$$

$$= y^2 \left(\frac{dx^2 + dy^2}{dx^2}\right), \text{ we have CE} = y \times \frac{\sqrt{dx^2 + dy^2}}{dx}$$

the expression for the normal.

Ex. Find the value of the normal and subnormal of a parabola.

$$y^2 = px, \text{ differentiating } 2ydy = pdx.$$

$$\frac{dy}{dx} = \frac{p}{2y}, \text{ mult. both sides by } y, \frac{ydy}{dx} = \frac{p}{2} \text{ subnormal.}$$

$$\text{Also, normal} = \sqrt{y^2 + \frac{p^2}{4}}.$$

SECTION V.

ASYMPTOTES TO CURVES.

To determine the method of drawing an asymptote to a curve.

(1.) If from a point in the axis of a curve at

any definite distance from the vertex, a straight line be drawn in such a direction that the curve continually approaches it, and may be made to approach it nearer than by any finite distance, but never meets it, it is called an *Asymptote* to the curve.

Let CG be an asymptote to the curve BA, and AE a tangent at the point A. Let the point A be taken farther and farther from B, then it is obvious that the subtangent DB will increase and the point E approach to C, and may be made to approach nearer to it than by any finite magnitude. Hence the asymptote CG is the *limit* of the tangent when BD becomes infinite. Hence, if we find the value of the subtangent, and consequently of BE, which, when x and y become infinite, is the same as BC; and also determine the point H from the triangles EAD, CHB, which are ultimately similar, we get the direction CH of the asymptote.

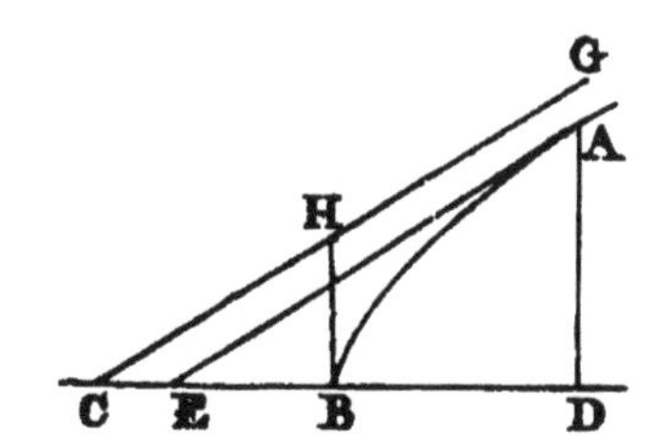

Ex. Let the curve be the hyperbola.

The subtangent will be found to be $\frac{2ax+x^2}{a+x}$ and BE —

$\frac{2ax+x^2}{a+x}-x=\frac{ax}{a+x}$ by reduction. Now, when x is

infinite, $a+x=x$, and hence the limit is $\frac{ax}{x}=a=$ half the transverse axis.

Hence, by similar triangles, DE : DA :: BC : BH.

That is, $\frac{2ax+x^2}{a+x} : \frac{b}{a}\sqrt{2ax+x^2} :: a :$ BH. When x is infinite the limit of the first term is x, and of the second $\frac{b}{a}\times x$. Hence $x : \frac{b}{a}\times x :: a :$ BH $= b$, half the conjugate axis. Hence, draw BH at right angles to the transverse axis, make it equal to the conjugate axis; join C, the centre of the hyperbola, or middle of the transverse axis, and the point H, and the line CH, produced indefinitely, will be the asymptote required.

NOTE. When the asymptotes CG, CG′, are at right angles to one another, the hyperbola is called a *Rectangular Hyperbola.*

SECTION VI.

LENGTH OF ARCS.

(1.) To find an expression for the length of an arc of a curve.

Let the ordinate (fig. p. 65) be supposed to move *uniformly* along the axis, whilst its other extremity describes the curve. Then the rate of increase of the abscissa being uniform, x will be the independent variable. The direction in which the extremity of the ordinate is moving when it reaches the point E, is in the tangent, and if the rate at which it was moving had continued uniform without any change in the direction of its motion, it would have moved uniformly along EH, whilst the ordinate was moving from the position ED to HK. Hence EH will represent the differential of the curve at the point E, or in other words, the rate at which the arc AE was increasing, when the point describing the arc was just passing the point E. Hence, the sides of this remarkable triangle EGH will represent the differentials of the abscissa, ordinate, and arc AE. Let the arc be denoted by z, then EG$=dx$, GH$=dy$, EH$=dz$. Hence $dz^2 = dx^2+dy^2$, and $dz = \sqrt{dx^2+dy^2}$.

In applying this formula to the arc of any curve, we obtain, from the equation of the curve, an expression equal to $\sqrt{dx^2+dy^2}$, which being the *differential* of the arc AE, its *integral* will be the real length of the arc itself, supposing it were unbent into a straight line. Hence the arc is said to be *rectified*, or made equal to a straight line, and the mode of doing so is called the *rectification* of the arc.

(2.) Before applying this formula to any particular curve, we shall first determine the rate at which the arc of a circle is increasing at E, by considering in succession the trigonometrical lines belonging to the arc, as the independent variable.

Let a denote the radius; AD or the versed sine $=x$; DE or sine $=y$; EF or tangent $=t$; and CF or secant $=s$. Then by the equation of the circle $y - \sqrt{2ax-x^2}$, DC or cosine of the arc $- \sqrt{a^2-y^2}$, and $t = \sqrt{s^2-a^2}$.

1. Let the ordinate move uniformly along the axis, then x increases uniformly, or CD diminishes uniformly; hence dx is the differential of the versed sine.

Hence, by similar triangles, $y : a :: dx : dz$. And

$$dz = \frac{adx}{y} = \frac{adx}{\sqrt{2ax-x^2}} = \frac{dx}{\sqrt{2-x^2}} \text{ when } a=1, \quad (1)$$

which is the differential of the arc AE, when the versed sine is the independent variable.

2. Let the ordinate or sine move along the axis so as to increase *uniformly*, or in other words, let y be the independent variable. Then

$$\sqrt{a^2-y^2} : a :: dy : dz.$$

Hence $dz = \dfrac{ady}{\sqrt{a^2-y^2}} = \dfrac{dy}{\sqrt{1-y^2}}$, when $a=1$ . . (2.)

(3). Let the secant increase uniformly, or be the independent variable.

Then $s : a :: a : \text{CD}-\dfrac{a^2}{s}$, hence $\text{AD} = a-\dfrac{a^2}{s}$.

Now the differential of $a-\dfrac{a^2}{s} = \dfrac{a^2ds}{s^2}$ by the rule for fractions:—

But $\sqrt{s^2-a^2} : s :: dx : dz :: \dfrac{a^2ds}{s^2} : dz$

Hence,

$$dz = \frac{a^2ds}{s\sqrt{s^2-a^2}} = \frac{ds}{s\sqrt{s^2-1}} \text{ when } a=1 \ldots \ldots (3)$$

(4.) Let the tangent increase uniformly, or become the independent variable.

Then $t = \sqrt{s^2-a^2}$ and $dt = \dfrac{sds}{\sqrt{s^2-a^2}}$

substituting dt instead of this value in equation . . (3.)

$$dz = \frac{a^2dt}{a^2+t^2} = \frac{dt}{1+t^2}, \text{ when } a - 1 \text{ . . . . (4.)}$$

APPLICATION OF THE LAST FORMULA.

To find the length of the arc of a circle.

Let a point move *uniformly* along the tangent from the point of contact, and let a straight line joining the centre and this point be supposed to move along with it, then this line will pass over the arc in such a manner, that whilst the tangent increases uniformly the arc does not increase uniformly. When the arc is equal to z, the rate of its increase was found to be expressed by $\frac{dt}{1+t^2}$.

$$\text{Since } dz = \frac{dt}{1+t^2} = dt \times \frac{1}{1+t^2}$$

$$\text{And } \frac{1}{1+t^2} = 1-t^2+t^4-t^6+, \text{ \&c.}$$

$$dz = \frac{dt}{1+t^2} = dt-t^2dt+t^4dt-t^6dt+, \text{ \&c.}$$

Hence $\int dz$ or $z = \int dt - \int t^2dt + \int t^4dt - \int t^6dt +$, &c.

$$\text{Or } z - 1 - \frac{t^3}{3} + \frac{t^5}{5} - \frac{t^7}{7} + \frac{t^9}{9} -, \text{ \&c.}$$

If we take an arc of 45^0, $t = r = 1$.

$$\text{Hence } z = 1 - \frac{1}{3} + \frac{1}{5} - \frac{1}{7} + \frac{1}{9} -, \text{ \&c.}$$

This series diminishes very slowly, and, consequently, would require a great number of terms to give a good approximation to the length of the arc.

If we take the arc $= 30^{\circ}$, then ED $= .5$.

And CD being found, we have the following proportion:

$$\text{CD} : \text{DE} :: \text{CA} : t = .5773502.$$

Hence, substituting this value for t in the above series, we have $z = .5235987$, which is the $\frac{1}{12}$ of the whole circumference. Hence, this number multiplied by 12 gives 6.2831804 the circumference of a circle whose radius is 1, and the half of this, or 3.14159023 &c. the circumference of a circle whose diameter is 1.

EXERCISES.

1. If two bodies start from the extremity of the diameter of a circle, the one moving uniformly along the diameter at the rate of 100 feet per second, and the other in the circumference, with a varying motion, so as to keep it always perpendicularly above the other; it is required to find its velocity in the circumference, when passing the sixtieth degree from the starting point, supposing the diameter of the circle 1000 feet?

2. If two bodies start from the extremity of the diameter of a circle at the same instant, the one moving uniformly along the tangent, at the rate of 100 feet per second, and the other along the circumference, with a variable motion, so as to be always in a straight line joining the first body, and the centre of the circle; it is required to find its velocity in passing the 45th degree from the starting point, the diameter being 2000 feet?

OBS. This section will give the pupil an idea of the principles by which Astronomers are enabled to determine the velocity of a planet in any point of its orbit. He will also perceive the reason of an experiment he may have frequently performed, viz. in throwing a stone by means of a sling, the stone moves off in a tangent to the circumference of the circle, at the point in which it was when he let go one end of the sling.

SECTION VII.

AREAS OF SURFACES.

To find a rule for obtaining the area of a curvilinear surface.

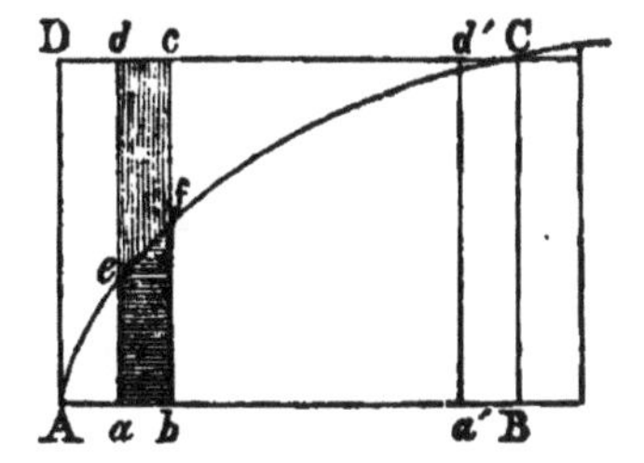

Let ABC be a surface bounded by the straight lines AB, BC, and the arc of a curve AC, and let an ordinate begin to move from A at a uniform rate along AB, it will describe or pass over the area A*f*CB. Construct the rectangle ABCD, and suppose AB divided into any number of equal parts, then one of these parts will represent the differential of AB or dx. Let the line AD move uniformly along with the ordinate, and it will describe the rectangle ABCD. Now, since the area of the rectangle increases uniformly, its corresponding increment or AD $\times dx$ will represent the differential of the area. The increments of the curvilinear surface are at first very small compared with the simultaneous increments of the rectangle, but rapidly approach to a ratio of equality as the ordinate approaches BC. When the ordinate has passed BC, the increments of the curvilinear surface are *greater* than the corresponding increments of the rectangle; consequently, the rate at which the curvilinear surface is increasing when the ordinate is passing BC, is the same with that of the rectangle. Hence the differential of the surface A*f*CB is equal to ydx. To apply this to any particular curve, we get an expression equal to dx, by differentiating the equation of the curve, and then for ydx, the integral of which gives the area required.

EXAMPLE 1.

To find the area of a parabola.

Since $y^2 = px$, $2ydy = pdx$, by differentiating;

Hence $dx = \frac{2ydy}{p}$ and $ydx = \frac{2y^2dy}{p}$, by mult. by y.

Therefore $\int\frac{2ydy}{p} = \frac{2y^3}{3p} = \frac{2}{3}yx$, by substituting the value of p.

That is, the parabola is $\frac{2}{3}$ of its circumscribing rectangle.

EXERCISES.

1. Required a rule for finding the area of a right-angled triangle, (fig. page 9) whose base AB is x, and perpendicular ax; or if BC be called y, the equation is $y=ax$?

2. Required a rule for finding the area of the curvilinear surface A*f*CD (last fig.) contained between the convex side of the parabola and the straight lines AD, DC, by supposing a line or ordinate to move from A uniformly along AD, one of the extremities keeping on the line AD, and the other in the curve?

EXAMPLE 2.

To find the area of a circle.

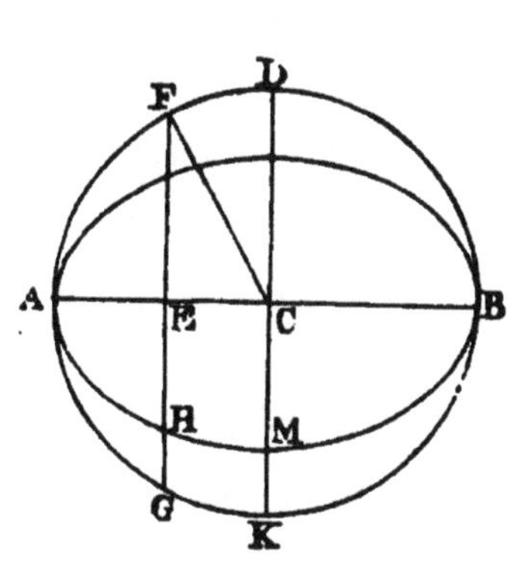

Let a line perpendicular to CA move from the position CD uniformly along CA, one of its extremities describing the arc DF. Let CE $= x$, EF $= y$, and the radius of the circle $= 1$. Then $y^2 = 1^2 - x^2$, and $y = \sqrt{1-x^2}$ or $(1-x^2)^{\frac{1}{2}}$, $y = (1-x^2)^{\frac{1}{2}} =$

$$1 - \frac{x^2}{2} - \frac{x^4}{8} - \frac{x^6}{16} - \frac{x^8}{128} -, \text{ \&c. by the Binom. Theor.}$$

Hence $ydx = dx - \dfrac{x^2dx}{2} - \dfrac{x^4dx}{8} - \dfrac{x^6dx}{16} - \dfrac{x^8dx}{128}, -$ &c.

Therefore

$$\int ydx = \int\left(dx - \frac{x^2dx}{2} - \frac{x^4dx}{8} - \frac{x^6dx}{16} - \text{\&c.}\right)$$

Hence,

area of ECDF $= x - \dfrac{x^3}{6} - \dfrac{x^5}{40} - \dfrac{x^7}{112} - \dfrac{5x^9}{1152} -$ &c.

by integrating.

If the arc DF $= 30°$, the sine of $30°$, or its equal CE, is half the radius, or $\frac{1}{2}$. Hence the series becomes $\frac{1}{2} - \frac{1}{48} - \frac{1}{1280} - \frac{1}{14336} -$ &c. If these be reduced to decimal fractions, and a sufficient number of terms subtracted from $\frac{1}{2}$ or .5, the result will be .4783055, which is the area of ECDF. If from this we subtract the area of the right-angled triangle ECF, which is .2165063, the remainder, or .2617992, will be the area of the sector FCD, which is the $\frac{1}{12}$ of the circle: Hence, 12 times this number, or 3.14159, &c.

is the area of a circle whose radius is 1. Now, as circles are as the squares of their diameters, $\frac{1}{4}$ of this, or .78539, &c. is the area of a circle whose diameter is 1.

COR. If an ellipse be described on AB as its transverse axis, then the transverse axis will be to the conjugate as the area of the circle to the area of the ellipse.

Let an ordinate be supposed to move from A uniformly along AB, then the variable ordinate EG will pass over the semicircle, and the variable ordinate EH will pass over half the ellipse. But since CK : CM :: EG : EH, that is, since the transverse is to the conjugate axis, as any ordinate of the circle to the corresponding ordinate of the ellipse, the areas passed over by these ordinates must be in the same ratio.

Ex. What is the area of an ellipse whose transverse axis is 20 inches and conjugate 10 inches?

SECTION VIII.

SURFACES OF SOLIDS.

To find the surfaces of solids of revolution.

Let a solid be generated by the revolution of the curvilinear surface AEB, about its axis AB. Let a circle whose plane is at right angles to AB begin to move from A uniformly along AB, its centre being in AB and

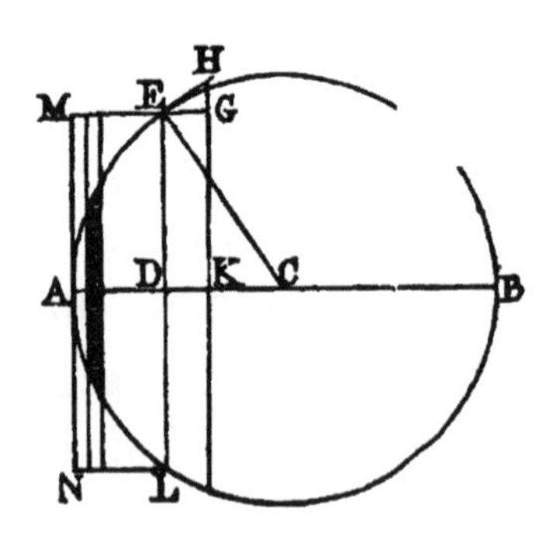

its *variable* circumference on the surface of the solid, then it is obvious that this variable circumference will pass over the entire surface of the solid. Now the rate of increase of the arc AE at the point E was formerly shown to be represented by EH, the corresponding or cotemporary rates of the ordinate and abscissa being represented by GH and EG. Hence, the rate at which the surface of the solid is increasing when the diameter of the generating circle is in EL, will be represented by the circumference of the circle multiplied by the rate of its motion in the direction of the curve at the point E, that is, by the circumference of the circle multiplied by EH. Let the radius of the circle or ED $= y$, and $\pi = 3.14159$, and arc AE $= z$, then the circumference is $2\pi y$, and the rate of increase or differential of the surface of the solid, $2\pi y dz$. Hence, from the equation of the curve we determine an expression equal to $2\pi y dz$, the *integral* of which will give the formula or rule for the surface of the solid.

EXAMPLE.

To find the surface of a sphere whose radius is r.

Since CE : DE :: EH : EG by similar triangles;

that is, $r : y :: dz : dx$, we have

$$ydz = rdx, \text{ and } 2\pi y dz = 2\pi r dx.$$

Hence $2\pi r dx$ is the *differential* of the surface of the segment when the plane of the variable circle is passing EL, or it is the rate at which the surface of the segment EAL is increasing; consequently, the *integral* of this will be the surface of the segment. That is,

$$\int 2\pi r dx = 2\pi r x = \text{surface of the segment.}$$

Now, as this is true whatever be the value of x,

let $x = r$, then $2\pi r^2$ — surface of the hemisphere,
and $4\pi r^2$ = surface of the sphere.

But πr^2 = area of a circle whose radius is r.
Hence the surface of a sphere is equal to *four* times the area of one of its great circles.

COR. 1. Since $2\pi rx$ is the expression for the convex surface of a cylinder whose radius is r and height x, it is obvious that the convex surface of a sphere is equal to the convex surface of its circumscribing cylinder.

COR. 2. Hence, if the sphere and its circumscribing cylinder be cut by a plane, the surface of the *segment* cut off will be equal to the convex surface of the corresponding segment of the cylinder.

COR. 3. If the sphere and its circumscribing cylinder be cut by two parallel planes, the surfaces of the corresponding *zones* are equal.

EXERCISES.

1. If the diameter of the earth be 8000 miles, how many square miles are contained in the Frigid Zone, and also, how many in the North Temperate Zone?

2. Let a variable circle move from the vertex of a cone and describe the conical surface, it is required to determine a rule for finding the convex surface?

SECTION IX.

CAPACITIES OF SOLIDS.

To find the capacities of solids of revolution.

Let the solid be generated by the revolution of the surface AEB (see last fig.), about its axis AB. Let a circle whose radius is AD, and whose plane is perpendicular to AB, move uniformly along AB, then it is obvious that the plane of this circle will generate a cylinder. Let a variable circle move uniformly along with the circle which describes the cylinder, its circumference describing the curved surface of the solid; then it is obvious, that the increments of the solid will at first be very *small* compared with the cotemporary increments of the cylinder, and that they will rapidly increase and tend to a ratio of equality as the describing circle approaches ED, and may be made to differ from each other by a quantity less than any finite magnitude. When the plane has passed the position ED, the increments of the solid will exceed the corresponding increments of the cylinder. Hence, as the cylinder increases uniformly, its increment generated in any portion of time may be taken to represent its differential, and as the solid increases at the same rate as the cylinder when the plane of the generating circle is passing ED, the differential of the solid is the same as the differential of the cylinder. Now, if AD be divided into any number of equal parts, one of these, which we may represent by dx, will be the differential of AD, and the area of the generating circle multiplied by dx, the differential of the cylinder. Let a =.7854, b = diameter of the circle or EL, then ab^2dx = differential of the solid. To apply this to any particular case, we obtain an expression equal to ab^2dx, the integral of which will be the solidity required.

EXAMPLE.

To find the solidity of a sphere.

Let r = the radius of the sphere, then

$$y^2 = (2r - x)x = 2rx - x^2, \text{ but } b = 2y.$$

Hence $b^2 = 8rx - 4x^2$. Multiplying both sides by adx we have $ab^2dx = 8rxdx - 4ax^2dx$. Hence,

$\int(8arxdx - 4ax^2dx) = 4arx^2 - \frac{4}{3}ax^3 =$ solidity of the segment EAL.

Let $x = r$, then $4ar^3 - \frac{4}{3}ar^3 =$ solidity of the hemisphere, and $8ar^3 - \frac{8}{3}ar^3$ of the sphere If D be the diameter $aD^3 - \frac{1}{3}aD^3 = \frac{2}{3}aD^2 \times D = \frac{2}{3}$ of the circumscribing cylinder.

COR. If an ellipse be described on AB as the major axis, and if two circles move from A uniformly along AB, one of them describing the sphere and the other the spheroid, the radii of these circles, and consequently their diameters, are in every position in the constant ratio of the transverse to the conjugate axis, and their areas as the squares of these axes. Hence the solidities passed over by these circles will be in the same ratio; that is, the square of the transverse is to the square of the conjugate, as the solidity of the sphere to the solidity of the spheroid. Let a — transverse and b the conjugate axis, then

$a^2 : b^2 :: \frac{2}{3} \times .7854 \times a^3 : \frac{2}{3} \times .7854 \times b^2 \times a = \frac{2}{3}$ of the cylinder, whose diameter is b and height a. Hence the spheroid is $\frac{2}{3}$ of its circumscribing cylinder.

EXERCISES.

1. If the diameter of the earth be 8000 miles, how many cubic miles are contained in the segment cut off by a plane passing through the Arctic Circle?

2. Find by the same principles a rule for the solidity of a cone.

3. Find a rule for determining the solidity of parabolic conoid, or solid formed by the revolution of a parabola about its axis.

OBSERVATIONS ON THE PRECEDING PART.

The learner generally finds some difficulty in comprehending how it happens, that whilst the independent variable is the same in algebraic expressions, which are not equal to each other, these expressions or functions have nevertheless *equal* differentials. The difficulty will vanish if he reflect, that the rate of increase does not depend on the magnitude of the function, but on the manner in which the independent variable enters into the expression. An expression whose value is small may notwithstanding increase as rapidly as one whose value is great. Thus, if we have two rectangles, whose altitudes are equal, but whose bases are unequal, and if the differential of the base or dx, be the same in both, the differential of these unequal rectangles are both adx. Take, as another example, a triangle whose base is x and perpendicular y; and a parabola whose abscissa is x, and ordinate y. Then the equation of the straight line or hypothenuse of the triangle, is $y = ax$, and that of the curve $y^2 = px$.

Differentiating the first equation, we have $dy = adx$, and multiplying both sides by y, and dividing by a, we have $ydx = \frac{ydy}{a}$ which is the differential of the area of the triangle.

Again, differentiating both sides of the equation of the parabola, we have

$2ydy = pdx$, and multiplying by y and dividing by p,

$$ydx = \frac{2y^2dy.}{p} \quad \text{Hence } \frac{ydy}{a} = \frac{2y^2dy}{p},$$

two equal differentials resulting from unequal functions. Hence, though these differentials be equal, their *forms* are different, which shows that they have been derived from different functions.

The learner will now clearly see that two equal differentials when integrated may give two unequal integrals. Thus, the integral of the first is

$\int\frac{ydy}{a} = \frac{y^2}{2a}$, which is the area of the triangle whose base is $\frac{y}{a}$ or x, and perpendicular y; and $\int\frac{2y^2dy}{p} = \frac{2}{3}\frac{y^3}{p} = \frac{2}{3}\frac{y^2}{p} \times y$, which is the area of a parabola whose ordinate is y, and abscissa $\frac{y^2}{p}$, or x.

In both these integrals the constant is 0, and yet they are unequal, and as the value of the constant may be different in different integrals resulting from equal differentials, the learner will perceive that he can draw no conclusion with regard to the equality or inequality of integrals having equal differentials.

PROMISCUOUS QUESTIONS AND EXERCISES.

1. What do you understand by the maximum of a function?

2. What do you understand by the minimum of a function?

3. If $x+y=a$, can the value of a be determined?

4. If $x+y=a$, and $x \times y=$ a maximum, can the values of x and y be determined?

5. If an ordinate move uniformly along the axis of a curve, what do you understand by the differential of the ordinate when the abscissa becomes equal to a given quantity?

6. What do you understand by the differential of an arc of the curve, the circumstances being the same?

7. If a point move uniformly along a curve at a given rate, what do you understand by the differential of the abscissa and ordinate when the point has arrived at a certain distance from the vertex?

8. If a point move uniformly along the axis of a curve, and if you have given the rate of its motion, its distance from the vertex and the ratio of its motion to the rate of increase of the ordinate at that point; by what means would you ascertain the exact length of the ordinate?

9. When you speak of the direction of the motion of a point in a curve, what do you really understand by the expression, since no part of the curve is a straight line?

10. What is the general expression for the subtangent of a curve?

11. Describe the mode by which you determine

the rate at which the area of a curve is increasing, for a given ordinate and abscissa, and consequently the method of finding its area.

12. Describe the method of determining the area of the surface of a solid of revolution.

13. Do the same for the solidity.

14. In the applications of the differential calculus, do you ever *neglect* quantities which you consider *indefinitely* small; and if you do so, how do you know that the result you have obtained is the true result, or only an approximation to the truth?

15. If the base of a cone increase uniformly at the rate of 1 inch per second, at what rate is its solidity increasing when the diameter of the base becomes 10 inches, the height being constantly one foot?

16. At what rate does the solidity vary if the perpendicular diminish at the rate of 3 inches per second, and become equal to 18 inches, when the base, increasing at the rate of 1 inch per second, becomes equal to 10?

17. If the major axis of an ellipse increase uniformly at the rate of 2 inches per second, and the minor axis at the rate of 3, at what rate is the *area* increasing, when the major axis becomes 20 inches, the minor axis at the same instant being 12 inches?

18. If the ellipse in the last example be made to revolve about its major axis to form a spheroid, at what rate is the solidity of the spheroid increasing at the same moment, the data being the same as in the last example?

19. There is a remarkable curve, whose abscissa is x, and ordinate y, and the nature of the curve is such, that the uniform rate of increase of the abscissa is to

the rate at which the ordinate is increasing when the abscissa is x, as 1 is to $\frac{a-2x}{2\sqrt{2ax-x^2}}$: required, the equation of the curve, and consequently the curve itself?

20. There is another remarkable curve, whose abscissa is x, and is of such a nature that the rate of increase of the abscissa is to the rate of increase of the ordinate as 1 is to $\frac{\sqrt{a}}{2\sqrt{x}}$: required, the nature of the equation of the curve, and consequently the curve itself?

21. If a cone, a hemisphere, and cylinder stand on the same base, and have the same altitude; it is required to show that the *differentials* of these solids are *equal*, but that the *integrals* of these equal differentials are to one another as the numbers 1, 2, 3?

22. If $x+y=a$, and $x^2y^3=$ a maximum: required the values of x and y?

23. If $x+y+z=a$, and $xyz=$ a maximum: required the values of x, y, and z?

24. A London porter-brewer ordered his principal clerk to have a cylindrical vat constructed, the diameter of whose base, together with its height, should be 50 feet, and which should hold the *greatest* possible quantity of liquid: required, the diameter of the base, and height?

25. The same person ordered a rectangular trough to be constructed to hold 8000 cubic feet, and to require the *least* possible quantity of wood: required, its length, breadth, and depth?

26. A cabinet-maker has a mahogany board, the breadth at one end being 4 feet, and at the other 3, and its length 10 feet; and he wishes to cut the largest possible rectangular table out of it: at what distance from the narrow end must it be cut?

27. A carpenter has a tapering tree of valuable wood, the diameter of the larger end being 3 feet, and that of the lesser end a foot and a half, and the length 20 feet; and he wishes to cut the largest possible cylinder out of it: required the length and diameter of the cylinder?

28. A person applies to a turner to make him a cylinder which shall contain a cubic foot of wood, and have the least possible surface, including both ends: required, the diameter of the base and height?

PART III.

DEVELOPEMENT OF ALGEBRAIC EXPRESSIONS INTO INFINITE SERIES, DIFFERENTIATION OF TRANSCENDENTAL FUNCTIONS, AND INTEGRATION BY LOGARITHMS AND ARCS OF CIRCLES.

Before the learner can read this Part he ought to have studied plane trigonometry, the theory of indeterminate coefficients, and logarithms. But for the sake of those who may not have previously studied the two last, we shall give as much as is necessary to prevent any interruption by turning to other works on these subjects.

SECTION I.

INDETERMINATE COEFFICIENTS.

Let there be two infinite series arranged according to the ascending powers of x, having the first term independent of x, and the others with coefficients, which are also independent of x, then it may be shown that the coefficients of the same powers of x will be equal, provided the two series be equal to each other for every value of x.

Thus, if $a + bx + cx^2 + \&c. = A + Bx + Cx^2 + \&c.$

then $a = A$, $b = B$, $c = C$, &c.

For, let $x = 0$, then $a = A$, and $bx + cx^2 + \&c. = Bx + Cx^2 +$, and dividing by x, $b + cx + \&c. = B + Cx + \&c.$ Hence, when $x = 0$, $b = B$, &c.

Hence, if a fractional expression, or an expression having a fractional index, be put equal to such a series, and if the series be then multiplied by the denominator of the fraction, or in the case of a *surd*, raised to the corresponding power, the values of the quantities A, B, C, &c. may be determined by equating the coefficients of the same powers of x. Or, if the terms of the series be all brought to one side of the equation, with 0 on the other side, the coefficients of the same power of x will be equal to 0, from which the values of B, C, &c. will be found. If the developement of the given function should want any of the powers of x, that will be indicated in the result, by the coefficient of that term being 0.

EXAMPLE 1.

Required the developement of $\frac{a}{a-x}$?

Let $\frac{a}{a-x} = A + Bx + Cx^2 + Dx^3 + \&c.$ Then

$$a = a\text{A} + a\text{B}x + a\text{C}x^2 + a\text{D}x^3 + \&c.$$
$$\quad\quad\quad - \text{A}x - \text{B}x^2 - \text{C}x^3 - \&c.$$

Hence $a = a\text{A}$, and $\text{A} = \frac{a}{a} = 1$; $a\text{B} = \text{A}$, $\text{B} = \frac{1}{a}$;

$$a\text{C} = \text{B},\ \text{C} = \frac{1}{a^2};\ \text{D} = \frac{1}{a^3}\ \&c.$$

Hence $\frac{a}{a-x} = 1 + \frac{x}{a} + \frac{x^2}{a^2} + \frac{x^3}{a^3} + \&c.$

EXERCISES.

1. Required the developement of $\frac{1}{1+x}$?

2. Required the developement of $\frac{a}{a^2-x^2}$?

EXAMPLE 2.

Required the developement of $\sqrt{a-x}$ or $(a \quad x)^{\frac{1}{2}}$?

$$\text{Let } \sqrt{a-x} = \text{A} + \text{B}x + \text{C}x^2 + \text{D}x^3 + \&c.$$
$$\sqrt{a-x} = \text{A} + \text{B}x + \text{C}x^2 + \text{D}x^3 + \&c.$$

$$a - x = \text{A}^2 + \text{AB}x + \text{AC}x^2 + \text{AD}x^3 + \&c.$$
$$+ \text{AB}x + \text{B}^2x^2 + \text{BC}x^3 + \&c.$$
$$+ \text{AC}x^2 + \text{BC}x^3 + \&c.$$
$$+ \text{AD}x^3 + \&c.$$

Hence $a = \text{A}^2$ and $\text{A} = \sqrt{a} = a^{\frac{1}{2}}$; $-x = 2\text{AB}x$,

$$\text{or } -1 = 2\text{AB},\ \text{B} = -\frac{1}{2\text{A}} = -\frac{1}{2a^{\frac{1}{2}}};$$

consequently the coefficients of x, x^2, &c. in each of the remaining terms is $= 0$; $2\text{AC} + \text{B}^2 = 0$,

$\text{C} = \frac{-\text{B}^2}{2\text{AC}} = \frac{1}{4a} \div 2a^{\frac{1}{2}} = \frac{1}{8a^{\frac{3}{2}}}$; $2\text{AD} + 2\text{BC} = 0$,

$$D = \frac{2BC}{2A} = -\frac{-1}{2a^{\frac{1}{2}}} \times \frac{1}{8a^{\frac{3}{2}}} \div a^{\frac{1}{2}} = -\frac{1}{16a^{\frac{5}{2}}}; \ \&c.$$

Hence $\sqrt{a-x} = a^{\frac{1}{2}} - \frac{x}{2a^{\frac{1}{2}}} + \frac{x^2}{8a^{\frac{3}{2}}} - \frac{x^3}{16a^{\frac{5}{2}}} + \&c.$

EXERCISES.

1. Required the developement of $\sqrt{1-x}$?
2. Required the developement of $\sqrt{a^2+x^2}$?

SECTION II.

NATURE AND PROPERTIES OF LOGARITHMS.

(1.) Logarithms are a series of numbers in arithmetical progression corresponding to another series in geometrical progression. Thus,

0, 1, 2, 3, 4, 5, 6
1, 10, 100, 1000, 10000, 100000, 1000000.

The logarithm of 1 is 0; of 10, 1; of 100, 2; &c.

The term *logarithm*, derived from λόγων ἀριθμὸς (*logōn arithmos*), which denotes the *number of ratios*, is thus expressive of this property. For 1 : 10 :: 10 : 100; or, between 1 and 10 we have *one* ratio; between 1 and 100, *two* ratios of 1 to 10; between 1 and 1000, *three* ratios of 1 to 10, &c. Hence, the logarithm of all numbers between 1 and 10 will be a decimal fraction; of all numbers between 10 and 100, the logarithm will be 1, together with a decimal fraction, &c.

(2.) But though this was one of the earliest modes of considering logarithms, it is not the best for developing their remarkable properties. These will be most easily investigated by considering the logarithm of a number as the power to which a constant quantity, a, must be raised so as to be equal to a given number. Thus, in the equation $a^x = n$, x is the logarithm of n, the constant number a being called the *base* of the system.

(3.) The sum of the logarithm of two numbers is equal to the logarithm of their product.

Let $a^x = n$ and $a^{x'} = n'$, then $a^x \times a^{x'} - nn'$; that is, $a^{x+x'} = nn'$, or $x + x'$ is the log. of $n \times n'$.

(4.) The difference of the logarithms of two numbers is equal to the logarithm of the quotient of these numbers.

Thus, $a^x \div a^{x'}$, or $a^{x-x'} = \frac{n}{n'}$; that is,

$$x - x' \text{ is the logarithm of } \frac{n}{n'}$$

(5.) The logarithm of the nth power of a number is equal to n times the logarithm of the number.

Since $a^x = N$, $(a^x)^n = N^n$; that is, $a^{nx} = N^n$.

Hence nx = logarithm of N^n.

(6.) The logarithm of the nth root of a number is equal to the nth part of the logarithm of the number.

Since $a^x = N$, $\sqrt[n]{a^x} = \sqrt[n]{N}$, or $a^{\frac{x}{n}} = N^{\frac{1}{n}}$.

Hence $\frac{x}{n}$ is the logarithm of $\sqrt[n]{N}$.

NOTE. The learner must now acquire the habit of performing calculations by means of logarithms with facility and accuracy.

EXPONENTIAL THEOREM.

(7.) It is required to develope a^x into an infinite series.

Let $a = 1+a-1$, and let $a-1 = b$, then $a^x =$

$$(1+b)^x = 1+xb+x.\frac{x-1}{2}b^2+x.\frac{x-1}{2}.\frac{x-2}{3}b^3+\&c.$$

If this series be arranged according to the ascending powers of x, it will become

$1 + (b-\frac{1}{2}b^2+\frac{1}{3}b^3-\frac{1}{4}b^4+\&c.)\,x+\text{B}x^2+\text{C}x^3+$ &c. in which the coefficients B, C, &c. do not contain x.

The terms inclosed in the parentheses are found as follows:

$$x.\frac{x-1}{2}b^2 = \frac{x^2-x}{2}b^2 = \frac{x^2b^2}{2} - \frac{x}{2}b^2 = \tfrac{1}{2}b^2x^2 - \tfrac{1}{2}b^2x.$$

Hence the coefficient of x, derived from this term is $-\frac{1}{2}b^2$. In like manner, its coefficient from the next term will be $\frac{1}{3}b^3$, &c.

Let $a-1$ be substituted for b in the quantity inclosed in the parentheses, and let the whole be denoted by A, then

$$A = (a-1) - \tfrac{1}{2}(a-1)^2 + \tfrac{1}{3}(a-1)^3 - \&c.$$

Hence $a^x = 1 + Ax + Bx^2 + Cx^3 + \&c. \; . \; . \; . \; . \; (1).$

Also $a^y = 1 + Ay + By^2 + Cy^3 + \&c.$

$$\begin{aligned} a^x \times a^y = a^{x+y} = 1 + Ax + &Bx^2 + Cx^3 + \&c. \\ &Ay + A^2xy + ABx^2y + \&c. \\ &+ By^2 + ABxy^2 + \&c. \\ &+ Cy^3 + \&c. \end{aligned}$$

by multiplying together the two sides of these equations.

Again, since $a^{x+y} = 1 + A(x+y) + B(x+y)^2 + C(x+y)^3 + \&c.$

This series is equal to the preceding, and by the theory of indeterminate coefficients, the coefficients of the corresponding powers of x and $(x+y)$ must be equal.

Hence, since the first two terms are identical in both series, we have

$$Bx^2 + A^2xy + By^2 = B(x+y)^2 = Bx^2 + 2Bxy + By^2.$$

Hence $B = \frac{A^2}{2}$.

Again, $Cx^3 + ABx^2y + ABxy^2 + Cy^3 = C(x+y)^3 = Cx^3 + 3Cx^2y + 3Cxy^2 + Cy^3$.

Hence, $C - \frac{A^3}{2.3}$.

In like manner $D = \frac{A^4}{2.3.4}$, &c.

Substituting these values in equation (1),

$$a^x - 1 + \frac{Ax}{1} + \frac{A^2x^2}{1.2} + \frac{A^3x^3}{1.2.3} + \frac{A^4x^4}{1.2.3.4} + \&c.$$

which is the *Exponential Theorem.*

(8.) Let $x = 1$, then $a = 1 + \frac{A}{1} + \frac{A^2}{1.2} + \frac{A^3}{1.2.3} +$ &c.

Again, let $A = 1$, then

$$1 + \frac{1}{1} + \frac{1}{1.2} + \frac{1}{1.2.3} + \frac{1}{1.2.3.4} + \text{\&c.}$$

$= 2.71828$, by adding together a sufficient number of terms.

Let $e = 2.71828$.

Then $e^x = 1 + \frac{x}{1} + \frac{x^2}{1.2} + \frac{x^3}{1.2.3} +$ &c.

And $e^A = 1 + \frac{A}{1} + \frac{A^2}{1.2} + \frac{A^3}{1.2.3} +$ &c.

Consequently $e^A = a$, and A is the logarithm of a in the system whose base is 2.71828. This is called the Naperian system from the name of its illustrious inventor. These logarithms are frequently, though improperly, called Hyperbolic Logarithms. The base of the common system is 10, and is generally denoted by a.

We shall denote common logarithms by the contraction LOG. and Naperian logarithms by log.

COR. 1. Since $a^x = 1 + \frac{Ax}{1} + \frac{A^2x^2}{1.2} +$, &c.

$$a^{\frac{1}{A}} = 1 + \frac{1}{1} + \frac{1}{1.2} + \frac{1}{1.2.3} +, \text{\&c.}$$

Hence $a^{\frac{1}{A}} = e$, or taking the Naperian logarithms of both sides, we have

$\frac{1}{A} \times \log. a = \log. e = 1$, consequently $\frac{1}{A} = \frac{1}{\log. a.}$

The quantity $\frac{1}{\ }$, or its equal $\frac{1}{\log.\ a}$, is called the *Modulus* of the system, and is generally represented by M.

Hence $\frac{1}{\log.\ 10} = \frac{1}{2.30258509} - 434294482$,

which is the modulus of the common system, the number 2.30258509 being the Naperian logarithm of 10.

Cor. 2. Let $a^x = e^y = n$, then taking the Naperian logarithms of both sides, $x \log.\ a = y \log.\ e$.

Hence, $x : y :: \frac{1}{\log.\ a} : \frac{1}{\log.\ e}$, or 1.

That is, the logarithms of the same number in the common and Naperian systems are to each other as the *moduli* of the two systems. Hence, to convert common logarithms into Naperian, multiply by 2.30258509; and to convert Naperian into common, multiply by .434294482.

If the pupil multiply common logarithms by 2.3, he will obtain Naperian logarithms sufficiently accurate for performing the exercises, when used merely as *exercises* for practice.

SECTION III.

ON THE DIFFERENTIATION OF EXPONENTIAL AND LOGARITHMIC FUNCTIONS AND INTEGRATIONS.

(9.) To find the differential of a^x.

Let x increase uniformly, and become $x+h$, then

the new state of the function will be denoted by a^{x+h} which is equal to $a^x \times a^h$.

Now $a^h = 1 + \text{A}h + \text{B}h^2 + \text{C}h^3 +$, &c. by the preceding articles.

Hence $a^x \times a^h$, or $a^{x+h} - a^x + \text{A}a^x h + \text{B}a^x h^2 + \text{C}a^x h^3 +$, &c. mult. by a^x.

Subtracting a^x, we have the increment of the function equal to $\text{A}a^x h + \text{B}a^x h^2 +$ &c. Now, had the function increased uniformly whilst x increased uniformly, we should have had, $dx : d.\ a^x :: h : \text{A}a^x h + \text{B}a^x h^2 +$, &c. $:: 1 : \text{A}a^x + \text{B}a^x h +$ &c.

But as this is not the case, we must take the limit of the ratio when $h = o$. Hence the limiting ratio is 1 to $\text{A}a^x$.

$$\text{Now, if } u = a^x\,,\quad \frac{du}{dx} = \text{A}a^x\,,$$

And $du = \text{A}a^x dx$, in which A is the Naperian log. of a.

Rule. Multiply the exponential a^x by the Naperian logarithm of the base, and by the differential of x, the product will be the differential a^x.

Thus, $d.\ a^x = \log. a \times a^x \times dx$.

Ex. If x increase uniformly at the rate of 1, at what rate is a^x increasing when $a = 10$ and $x - 2$?

$dx : d.\ a^x :: 1 ; \log. 10 \times 100 :: 1 : 230.258509$, or the function a^x is increasing 230.25 times faster than x.

(10.) If the natural numbers increase uniformly, to find the rate at which the logarithms are increasing when the natural numbers are passing a certain limit, or in other words, to find the differential of the logarithm of a given number.

Let $a^u = x$, that is u = Log. of x. Then

$$\text{A}a^u du = dx \text{ and } du = \frac{dx}{\text{A}a^u} = \frac{dx}{\text{A}x} = \frac{1}{\text{A}} \times \frac{dx}{x}$$

$$= \text{M} \times \frac{dx}{x} = \frac{dx}{x} \text{ when M} = 1. \quad \text{Also, } \frac{du}{dx} = \frac{1}{x}$$

Rule. Divide the differential of the given number, by the number itself, the quotient will be the differential of the Naperian logarithm of the number.

Ex. If the natural numbers increase uniformly at the rate of 1 per second, at what rate are the Naperian logarithms increasing when the natural number becomes 2?

$$dx : du :: 2 : 1 :: 1 : \tfrac{1}{2},$$

that is, the logarithms are only increasing half as fast as the natural numbers when the natural numbers are passing the limit 2.

GEOMETRICAL ILLUSTRATION.

1. In the straight line AD take any number of equidistant points, and draw perpendiculars at these points, and take these perpendiculars in geometrical progression,

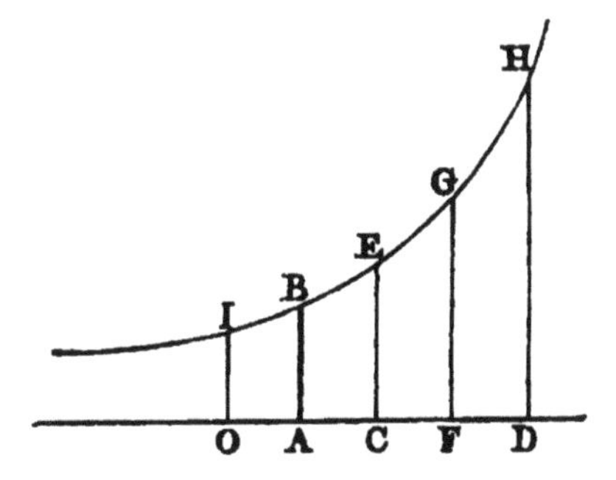

then the curve which passes through their extremities will be the logarithmic curve.

Let AB $= a$ the base, OI $= 1$,

Then, since $1 : a :: a :$ CE, CE $= a^2$,

And $a : a^2 :: a^2 :$ FG, FG $= a^3$.

Hence, OA is the log. of AB, OC the log. of CE, OF the log. of FG, &c.

2. Let an ordinate move *uniformly* along AB, that is, let the abscissa x reckoned from O, increase uniformly, required the rate at which the ordinate y is increasing when $x = 2$, the ratio of OI to AB being that of 1 to 10, or the curve being that which belongs to the common system

Since $a^x = y$, $d.\ a^x = dy$,

But $d.\ a^x = \log. a \times a^x \times dx$,

And $\dfrac{dy}{dx} = \log. 10 \times 10^2 = 230.2585093$,

Hence the ordinate CE is increasing 230.25 times faster than the abscissa. The ordinate FG is increasing 2302.58 times faster than the abscissa, &c.

3. Let the ordinate move along OB so as to increase uniformly, at what rate is the abscissa (reckoned from O) increasing when the ordinate becomes 100, the curve belonging to common logarithms?

$$a^u = x,\ \frac{du}{dx} = \text{M}\frac{1}{x} = .43429 \times \frac{1}{100} = .0043429, \&c.$$

That is, whilst the ordinate increases uniformly at the rate of 1, the abscissa is only increasing at the rate of .0043429 when the ordinate is equal to 100.

(11.) To find the differential of an exponential function of two independent variables, y^x.

Let $u = y^x$, taking the Naperian logarithm of both sides, log. $u = x$ log. y, differentiating both sides we have

$$\frac{du}{u} = x \times d. \log. y + \log. y \times dx, \text{ that is,}$$

$$\frac{du}{u} = \frac{xdy}{y} + \log. y. \ dx.$$

Hence $du = xy^{x-1}dy + \log. y. \ y^x. \ dx$.

Rule. Take the sum of the partial differentials by supposing first, y variable and x constant, and then x variable and y constant.

Cor. In like manner we may find the differential of y^{x^z}.

Let x^z be constant, and y variable, the partial differential will be $x^z y^{x^z - 1} dy$. Again, considering y constant, and x^z variable, we have the partial differential equal to $\log. y. \ y^{x^z} d. x^z$; but $d. x^z = zx^{z-1}dx + \log. x . x^z . dz$.

Hence $d. y^{x^z} - x^z y^{x^z - 1} dy + \log. y. \ y^{x^z}(zx^{z-1}dx + \log. x. x^z dx)$.

EXERCISES.

1. If y increase uniformly at the rate of 2, and x at the rate of 1, at what rate is the function y^x increasing at the moment y becomes 10, and x, 3?

2. If y increase uniformly at the rate of 1, x at the rate of 2, and z at the rate of 3, at what rate is the function y^{x^z} increasing when y becomes 10, x being equal to 4, and z equal to 5 at the same instant?

(12.) The integral of $\log. aa^x dx$ will obviously be a^x.

Thus, $\int \log. a a^x dx = a^x + c$.

Also the integral of $\frac{dx}{x}$ will be the log. of x.

Thus, $\int \frac{dx}{x} = \log. x + c$.

NOTE. If $\frac{dx}{x}$ be put into the form $x^{-1}dx$ it might appear to come under the form $x^{-n}dx$ in Part First. If we try to find its integral by the rule, for this form we should have $\int x^{-1}dx = \frac{x^0 dx}{0 \times dx} = \frac{x^0}{0} + c$ $= \frac{1}{0} + c =$ infinity, which shows that we are applying a rule to a particular example which does not come under that rule.

EXERCISES.

1. What is the integral of $\frac{2x dx}{a + x^2}$?

2. What is the integral of $\frac{dx}{x + a}$, and its numerical value when $x = 10$, a being 2?

3. What is the integral of $\frac{x^2 dx}{x^3}$?

4. What is the integral of $\frac{dx}{1 + x}$, and its numerical value, when $x = 1$?

(13.) The learner may, by combining the preceding rules with those in Part I. differentiate more complex exponential and logarithmic functions than those already given. We shall give him a few exercises, with some hints how to proceed.

EXERCISES.

1. Differentiate $\frac{a^x-1}{a^x+1}$, by the rule for fractions and that for exponential functions, and find the numerical rate of increase of this function when $a = 10$, and x becomes 2.

2. What is the differential of $(a^x+1)^2$, and at what rate is the function increasing when x becomes equal to 3, a being 2.71828?

3. What is the differential of $a^x \times a^y$, and at what rate is the function increasing when x increases at the rate of 2, y at the rate of 3, at the moment when x becomes equal to 1.5 and $y = 2.5$, the base a being 10?

4. Required the differential of the Naperian logarithm of $\frac{a+x}{a-x}$, and the rate of increase of the function log. $\frac{a+x}{a-x}$ when x increases uniformly at the rate of 1, becomes equal to 4, the constant a being 5?

5. Required the differential of $\left(\frac{a}{x}\right)^x$ or $\frac{a^x}{x^x}$?

$$d.\left(\frac{a}{x}\right)^x = \log.\frac{a}{x} \times \left(\frac{a}{x}\right)^x \times dx + x\left(\frac{a}{x}\right)^{x-1} \times d.\frac{a}{x}$$

The learner will reduce it to a simpler expression, and find the differential coefficient of the function $\left(\frac{a}{x}\right)^x$ when x becomes 2, the base a being 10.

(14.) When we meet with a differential in a fractional form, in which the numerator is the differential of the denominator, we have already seen that its integral is the Naperian logarithm of the denominator; but it frequently happens that we have to find the integral of fractional differentials, in which this is not the case, as the expression is given, but which may, by certain substitutions and other algebraic artifices, be brought to that form, and the integral obtained. The following are useful examples.

1st Form, $\frac{dx}{\sqrt{x^2 \pm a^2}}$.

Let $x^2 + a^2 = v^2$, then $2xdx = 2vdv$, or $xdx = vdv$. Hence $x : v :: dx : dv$, and $x+v : v :: dx+dv : dv$.

Therefore $\frac{dx+dv}{x+v} = \frac{dx}{v} = \frac{dx}{\sqrt{x^2+a^2}}$; but the first expression has its numerator, the differential of its denominator, and consequently $\int \frac{dx+dv}{x+v} = \log. (x+v)$;

therefore $\int \frac{dx}{\sqrt{x^2+a^2}} = \log. (x+\sqrt{x^2+a^2})$; in like manner, $\int \frac{dx}{\sqrt{x^2-a^2}} = \log. (x-\sqrt{x^2+a^2})$.

2nd Form, $\frac{dx}{\sqrt{x^2 \pm 2ax}}$.

Let $\sqrt{x^2+2ax} = v$, then $x^2 + 2ax = v^2$; and $x^2 + 2ax + a^2 = v^2 + a^2$, extracting the square root.

$x \pm a = \sqrt{v^2+a^2}$; hence $dx = d.(v^2+a^2)^{\frac{1}{2}} = \frac{vdv}{\sqrt{v^2+a^2}}$.

Hence $\frac{dx}{\sqrt{x^2+a^2}} = \frac{dv}{\sqrt{v^2+a^2}}$, by dividing the first side of the equation by $\sqrt{a^2+a^2}$, and the second by its equal v. Now, $\frac{dv}{\sqrt{v^2+a^2}}$ is the first form, which has already been integrated. Hence

$$\int \frac{dv}{\sqrt{v^2+a^2}} = \log.\,(v+\sqrt{v^2+a^2}) =$$

$$\log.\,(x+a+\sqrt{x^2+2ax}.)$$

Also, by taking $\sqrt{x^2-2ax} = v$, we obtain

$$\int \frac{dx}{\sqrt{x^2-2ax}} = \log.\,(x-a+\sqrt{x^2-2ax}).$$

3rd Form, $\frac{2adx}{a^2-x^2}$ or $\frac{2adx}{x^2-a^2}$.

Since $\frac{2adx}{a^2-x^2} = \frac{2adx}{(a+x)(a-x)} = \frac{dx}{a+x} - \frac{-dx}{a-x}$,

$$\int\left(\frac{dx}{a+x} - \frac{dx}{a-x}\right) = \int \frac{dx}{a+x} - \int \frac{-dx}{a-x} =$$

$$\log.\,(a+x) - \log.\,(a-x) = \log.\,\frac{a+x}{a-x}.$$

$$\text{Also } \int \frac{2adx}{x^2-a^2} = \log.\,\frac{x-a}{x+a}.$$

4th Form, $\frac{2adx}{x\sqrt{a^2+x^2}}$ or $\frac{2adx}{x\sqrt{a^2-x^2}}$

Let $\sqrt{a^2+x^2} = v$, then $a^2+x^2 = v^2$, and $xdx = vdv$.

Therefore $\frac{2adx}{xv} = \frac{2adv}{x^2}$, by substituting $\frac{xdx}{dv}$ for v.

Hence $\frac{2adx}{x\sqrt{a^2+x^2}} = \frac{2adv}{v^2-a^2}$, by substituting the value of v in the first side of the equation, and the value of x^2 or v^2-a^2 in the second. But the integral of the form $\frac{2adv}{v^2-a^2}$ is equal to log. $\frac{u-a}{u+a}$ by the last example.

Hence $\int\frac{2adx}{x\sqrt{a^2+x^2}} = \log.\frac{\sqrt{a^2+x^2}-a}{\sqrt{a^2+x^2}+a}$. In like manner, $\int\frac{2adx}{x\sqrt{a^2-x^2}} = \log.\frac{a-\sqrt{a^2-x^2}}{a+\sqrt{a^2-x^2}}$.

5th Form, $\frac{x^{-2}dx}{\sqrt{a^2+x^{-2}}}$.

Let $v = \frac{1}{x}$, then $dv = -x^{-2}dx$, and $-dv = x^{-2}dx$.

Hence $\frac{x^{-2}dx}{\sqrt{a^2+x^{-2}}} = \frac{-dv}{\sqrt{a^2+v^2}}$, which is the first form.

Therefore $\int\frac{-dv}{\sqrt{a^2+v^2}} = -\log.(v+\sqrt{a^2+v^2}) -$

$-\log.\left(\frac{1}{x}+\sqrt{a^2+\frac{1}{x^2}}\right) = -\log.\frac{1+\sqrt{1+a^2x^2}}{x}$.

Obs. We would advise the learner to collect into tables those *forms* of differentials whose integrals he has been able to obtain. Thus the results of the investigations in this section may be arranged as follows.

1. $\int\log. aa^xdx = a^x$.

2. $\int\frac{dx}{x} = \log. x$.

3. $\int x^{-1} dx = \log. x.$

4. $\int (xy^{x-1}dy + \log. y. y^x dx) = y^x.$

5. $\int \left\{ x^z y^{x^z-1} dy + \log. y. y^{x^z} (z x^{z-1} dx + \log. x. x^z dx) \right\}$
$= y^{x^z}$

6. $\int \frac{dx}{\sqrt{x^2 \pm a^2}} = \log. (x \pm \sqrt{x^2 \pm a^2}).$

7. $\int \frac{dx}{\sqrt{x^2 \pm 2ax}} = \log. (x + a + \sqrt{x^2 \pm 2ax}.$

8. $\int \frac{2adx}{a^2 - x^2} = \log. \frac{a+x}{a-x}.$

9. $\int \frac{2adx}{x^2 - a^2} = \log. \frac{x-a}{x+a}.$

10. $\int \frac{2adx}{x\sqrt{a^2 + x^2}} = \log. \frac{\sqrt{a^2 + x^2} - a}{\sqrt{a^2 + x^2} + a}.$

11. $\int \frac{2adx}{x\sqrt{a^2 - x^2}} = \log. \frac{a - \sqrt{a^2 - x^2}}{a + \sqrt{a^2 - x^2}}.$

12. $\int \frac{x^{-2}dx}{\sqrt{a^2 + x^{-2}}} = -\log. \frac{1 + \sqrt{1 + a^2x^2}}{x}.$

SECTION IV.

ON THE DIFFERENTIATION OF CIRCULAR FUNCTIONS AND INTEGRATION BY ARCS OF CIRCLES.

(1.) To find the differential of the arc of a circle.

Let AB be the diameter of a circle, then, if the tangent FEH be drawn, and EG at right angles to HK, it has already been shown (page 71) that if the arc AE

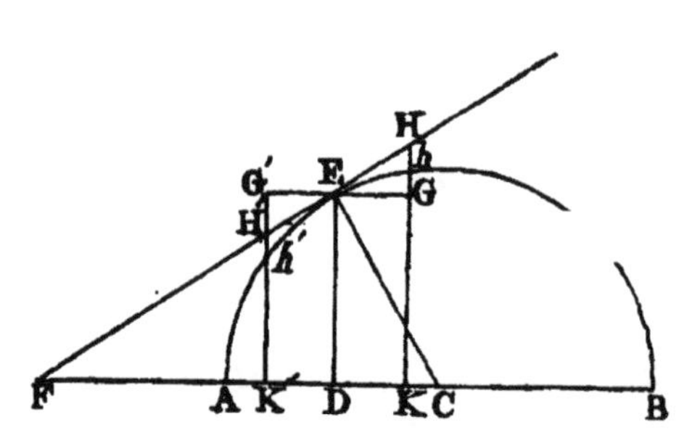

be represented by z, AD by x, and DE by y, EG $= dx$, GH $= dy$, and EH $= dz$. Hence we may easily determine, by means of the similar triangles EGH, EDC, FEC, the rate at which the arc AE is increasing or diminishing at the point E, the various trigonometrical lines being considered in succession as the independent variable.

Let a equal the radius; $x =$ the versed sine AD; $y =$ sine DE; $y' =$ cosine CD; $t =$ tangent EF; and $t' =$ cotangent; $s =$ secant CF; and s' — cosecant.

Then $y = \sqrt{2ax-x^2}$; $y' - \sqrt{a^2-y^2}$; $t = \sqrt{s^2-a^2}$.

1. Let the ordinate or sine move *uniformly* along AB, that is, let x be the independent variable.

Since DE : CE :: EG : EH,

Or, $y : a :: dx : dz$,

$$dz = \frac{adx}{y} = \frac{adx}{\sqrt{2ax-x^2}} \ldots\ldots (1)$$

2. Let the ordinate or sine move along AB so as to increase uniformly; that is, let y be the independent variable.

Since CD : CE :: GH : EH,

Or $y' : a :: dy : dz$,

$$dz = \frac{ady}{y'} = \frac{ady}{\sqrt{a^2-y^2}} \ldots\ldots (2)$$

3. Let the cosine be the independent variable. Then, since the arc *diminishes* when the cosine increases, at the same rate as it *increases* when the sine is the independent variable, the form of the differential will be the same as for the sine, except the change of sign from *plus* to *minus*. The same remark applies to the cotangent and cosecant.

$$\text{Hence, } dz = \frac{-ady'}{y} = \frac{-ady'}{\sqrt{a^2-y'^2}} \ldots\ldots (3)$$

4. Let the secant be the independent variable.

Since CF : CE :: CE : CD,

Or $s : a :: a : \text{CD} - \frac{a^2}{s}$

We have AD or $x = a - \frac{a^2}{s}$, and $dx = \frac{a^2ds}{s^2}$

Again FE : CF :: GH : EH,

Or $t : s :: dx : dz$,

That is, $t : s : \frac{a^2ds}{s^2} : dz$,

$$\text{Hence } dz = \frac{a^2ds}{st} = \frac{a^2ds}{s\sqrt{s^2-a^2}} \ldots\ldots (4)$$

5. Let the cosecant be the independent variable.

Then, $dz = \frac{-ads'}{s't'} = \frac{-a^2ds'}{s'\sqrt{s'^2-a^2}} \dots\dots(5)$

6. Let the tangent be the independent variable.

Then, since $t = \sqrt{s^2-a^2}$; $dt = \frac{sds}{\sqrt{s^2-a^2}}$

But $dz = \frac{a^2ds}{s\sqrt{s^2-a^2}} = \frac{a^2sds}{s^2\sqrt{s^2-a^2}}$, mult. both numerator and denominator by s. Hence, by substituting dt instead of $\frac{sds}{\sqrt{s^2-a^2}}$ in the last expression we have

$$dz = \frac{a^2dt'}{s^2} = \frac{a^2dt}{a^2+t^2} \dots\dots(6)$$

7. Let the cosecant be the independent variable.

$$\text{Then } dz = \frac{-a^2dt'}{s'^2} = \frac{-a^2dt'}{a^2+t'^2} \dots\dots(7)$$

EXERCISE.

If the versed sine, sine, tangent, and secant increase uniformly in succession at the rate of one inch per second, at what rate is the corresponding arc increasing when the versed sine is 50000, the sine 20000, the tangent 1000000, and the secant 1500000, the radius of the circle being 100000?

(2.) We are now enabled to integrate several new forms of differentials by means of the arcs of circles.

Thus, since $dz = \frac{ady}{\sqrt{a^2-y^2}}, \int dz$ or $z = \int \frac{ady}{\sqrt{a^2-y^2}}$.

Hence it is obvious that if we know the numerical values of the radius a and the sine y, we can easily determine the length of the arc z. The results obtained in the last division may be conveniently arranged as follows.

When radius $= a$.	When radius $= 1$.
1. $\int \frac{adx}{\sqrt{2ax-x^2}} = z$	2. $\int \frac{dx}{\sqrt{2x-x^2}} = z$.
3. $\int \frac{ady}{\sqrt{a^2-y^2}} = z$.	4. $\int \frac{dy}{\sqrt{1-y^2}} = z$.
5. $\int \frac{a^2dt}{a^2+t^2} = z$	6. $\int \frac{dt}{1+t^2} = z$.
7. $\int \frac{a^2ds}{s\sqrt{s^2-a^2}} = z$.	8. $\int \frac{ds}{s\sqrt{s-1}} = z$.

These integrals require no correction, for when x, y, t, or s, is 0, the arc z is 0, and consequently the constant is 0. If *minus* be placed before these expressions, we obtain the forms for the coversine, cosine, cotangent, and cosecant.

EXERCISES.

1. What is the value of the integral of $\frac{a^2dx}{a^2+x^2}$ when $a = 100000$, and $x = 100000$?

2. What is the numerical value of the integral of $\frac{adx}{\sqrt{2ax-x^2}}$ when a — 100000, and $x = 50000$?

3. What is the numerical value of the integral of $\frac{dx}{1-x^2}$, when $x = .5$?

4. What is the value of the integral of $\frac{dx}{x\sqrt{x-1}}$ when $x = 5$?

COR. 1. If the two first forms in the first column be divided by a, and the second pair by a^2, we shall obtain a new series of differentials whose integrals are known.

Thus, $\int \frac{adx}{\sqrt{2ax - x^2}} = za$; $\int \frac{dx}{\sqrt{2ax - x^2}} = z$.

And $\int \frac{dx}{\sqrt{2ax - x^2}} = \frac{1}{a}z$. Hence

1. $\int \frac{dx}{\sqrt{2ax - x^2}} = \frac{1}{a}z$; 2. $\int \frac{dy}{\sqrt{a^2 - y^2}} = \frac{1}{a}z$.

3. $\int \frac{dt}{a^2 + t^2} = \frac{1}{a^2}z$; 4. $\int \frac{ds}{s\sqrt{s^2 - a^2}} = \frac{1}{a^2}z$.

COR. 2. If we suppose the radius equal to $\frac{a}{b}$ in the *three* last expressions, or if we substitute $\frac{a}{b}$ for a, we shall obtain a new series of differentials, whose integrals will thus be determined.

Thus $\frac{dy}{\sqrt{a^2 - y^2}}$ becomes $\frac{dy}{\frac{1}{b}\sqrt{a^2 - b^2y^2}}$. Hence,

$$\int \frac{dy}{\frac{1}{b}\sqrt{a^2 - b^2y^2}} = \frac{1}{\frac{a}{b}}z = \frac{b}{a}z.$$

Also $b\int \frac{dy}{\sqrt{a^2 - b^2y^2}} = \frac{b}{a}z$ and $\int \frac{dy}{\sqrt{a^2 - b^2y^2}} = \frac{1}{a}z$.

By going through the same process with the remaining two we obtain the following results.

$$\int \frac{dt}{a^2 + b^2t^2} = \frac{1}{a^2}z; \quad \int \frac{ds}{s\sqrt{b^2s^2 - a^2}} = \frac{b}{a^2}z.$$

The sign *minus* placed before these will give the forms for the cosine, cotangent and cosecant, the radius being $= \frac{a}{b}$.

If we substitute $\frac{a^2}{2b^2}$ for a in the first expression, we obtain

$$\int \frac{dx}{\sqrt{a^2x - b^2x^2}} = 2\frac{b}{a^2}z, \text{ when the radius } = \frac{a^2}{2b^2}.$$

EXERCISES.

1. What is the value of $\int \frac{dx}{\sqrt{2ax - x^2}}$ when $a =$ 100000 and $x = 20000$?

2. What is the value of $\int \frac{dx}{x\sqrt{b^2x^2 - a^2}}$ when $a =$ 100000, $b = 150000$, and $x = 200000$?

3. What is the value of $\int_a^b \frac{dx}{1+x^2}$ between the limits $a = 2$ and $b = 3$?

(3.) To find the differentials of the sine, tangent, secant, &c. of an arc of a circle, the arc being the independent variable. These may easily be deduced from the differentials of the arc in preceding numbers.

Since $dz = \frac{adx}{y}$; $dx = \frac{1}{a}ydz$; that is,

$$d.\text{versin } z = \frac{1}{a}\sin zdz = \sin zdz, \text{ when } a = 1. \quad (1)$$

Since $dz = \frac{ady}{y'}$; $dy = \frac{1}{a}y'dz$; that is,

$$d.\sin z - \frac{1}{a}\cos z dz = \cos z dz, \text{ when } a = 1 \ldots (2)$$

Since $dz = \dfrac{-ady'}{y}$; $dy' = -\dfrac{1}{a}ydz$; that is,

$$d.\cos z = -\frac{1}{a}\sin z dz = -\sin z dz, \text{ when } a = 1. (3)$$

Since $dz = \dfrac{a^2 dt}{s^2}$; $dt = \dfrac{1}{a^2}s^2 dz$; that is,

$$d.\tan z = \frac{1}{a^2}(\sec z)^2 dz = (\sec z)^2 dz, \text{ when } a=1. . (4)$$

Since $dz = \dfrac{-a^2 dt'}{s'^2}$; $dt' = -\dfrac{1}{a^2}s'^2 dz$; that is,

$$d.\cot z = -\frac{1}{a^2}(\text{cosec } z)^2 dz = -(\text{cosec } z)^2 dz, \text{ if } a=1. (5)$$

Since $dz = \dfrac{a^2 ds}{st}$; $ds = stdz$; that is,

$$d.\sec z = \frac{1}{a^2}\sec z\tan z dz = \sec z\tan z dz, \text{ if } a=1. (6)$$

Since $dz = \dfrac{-a^2 ds'}{s't'}$; $ds' = -\dfrac{1}{a^2}s't'dz$; that is,

$$d.\cot z = -\frac{1}{a^2}\text{cosec } z\cot z dz = -\text{cosec } z\cot z dz,$$
$$\text{when } a = 1. \ldots\ldots\ldots (7)$$

NOTE. Had we determined the differentials of the sine, tangent, &c. from first principles, as is generally done, the differential of the *arc* would have been obtained from those of the sine, tangent, &c. Thus, if $d.\sin z = \cos z dz$, $dz = \dfrac{d.\sin z}{\cos z}$. Sir John Herschel, by considering the *arc* as the inverse function of the sine, tangent, &c. and using -1 to denote inverse, has proposed the following notation, which is chiefly adopted in this country. If $\sin x$ denote the *sine* of the *arc* x, then $\sin^{-1} x$ denotes the *arc* of the sine x, that is, the arc whose sine is x; and $d \sin^{-1} x$, the differential of the arc whose sine is x. In like manner, $\log.^{-1} x$ denotes the number whose logarithm is x, and $d \log.^{-1} x$ the differential of the number whose logarithm is x.

To accustom the learner to this notation he may write over the results in division (1) and (2) by using x as the independent variable. Thus,

$$dz = \frac{ax}{\sqrt{2ax - x^2}} \text{ is written, } d \operatorname{versin}^{-1} x = \frac{ax}{\sqrt{2ax - x^2}}.$$

Also,

$$\int \frac{ady}{\sqrt{a^2 - y^2}} = z, \text{ is expressed by, } \int \frac{adx}{\sqrt{a^2 - x^2}} = \sin^{-1} x.$$

EXERCISES.

1. If a point start from the extremity of the diameter of a circle, and move uniformly in the circumference at the rate of 1 inch per second, at what rate is the versed sine, sine, tangent, and secant increasing when the point is passing the *thirtieth* degree, the radius being 100000?

2. The data being the same, at what rate is coversed sine, cosine, cotangent, and cosecant diminishing, when the point is passing the sixtieth degree?

(4.) By the application of the preceding formulæ we may differentiate more complex forms than those which we have previously considered. We shall give a few examples.

1. Required the differential of the *sine of the sine of* z, that is, if the sine of z be bent into the arc of a circle, it is required to find the differential of the sine of that arc?

Let $u =$ sin of $\sin z$. Assume $\sin z = y$, then

$$u = \sin y\text{; and } du = \cos y dy = \cos y \cos z dz$$
$$= \cos \sin z \cos z dz.$$

2. The pupil will show that $d.$ sin of $\cos z =$
$$- (\cos z)^2 \sin z dz.$$

3. It is also required to show that $d. \sin \log. z =$
$$\cos \log. z \frac{dz}{z}.$$

4. If the sine of an arc of 30° be bent into the arc of a circle, and if a point move uniformly in the circumference at the rate of an inch per second, required the rate at which the sine of the sine of 30° is increasing, the radius being 100000?

5. At what rate is the sine of the arc equal to the Naperian logarithm of an arc of 30° increasing, the other data being the same?

6. What is the differential of $\sin z \times \cos z$?

7. What is the differential of $\frac{\sin z}{\cos z}$?

(5.) By means of the formulæ in division (3) we obtain the integrals of the following expressions, in which z, or the arc of a circle, is the independent variable.

When rad. $= a$.	When rad. $= 1$.
1. $\int \frac{1}{a} \sin z\, dz = \text{versin}\, z.$	2. $\int \sin z\, dz = \text{coversin}\, z.$
3. $\int \frac{1}{a} \cos z\, dz = \sin z.$	4. $\int \cos z\, dz = \sin z.$
5. $\int \frac{1}{a^2} (\sec z)^2 dz = \tan z.$	6. $\int (\sec z)^2 dz = \tan z.$
7. $\int \frac{1}{a^2} \sec z \tan z\, dz = \sec z$	8. $\int \sec z \tan z dz = \sec z.$

If *minus* be placed before each of these expressions we have the forms for the coversin, cosine, &c. Thus, $-\int \frac{1}{a} \sin z dz = \cos z$, which is left as an exercise for the learner.

EXERCISES.

1. What is the value of $\frac{1}{a} \int \cos z\, dz$ when $a -$ 100000 and $z = 50000$?

2. Find the numerical value of $\int_a^b \frac{1}{r^2} (\sec z)^2 dz$ when $r = 100000$, $a = 20°$, and $b = 30°$?

SECTION V.

ON THE NOTATION OF FUNCTIONS IN GENERAL, AND THEIR DEVELOPEMENT.

(1.) When an algebraic expression containing x is spoken of *generally*, without regard to its *particular* form, the letters f, F, ϕ, ψ, are used instead of the term *function*. Thus fx may denote x^2, x^n, $\sqrt{a+x}$, log. x, sin x, &c. When any function of x enters into an algebraic expression as a single letter, the expression is inclosed, and the letter f, F, &c. placed before it. Thus, (x^2+ax^2) may be viewed as a function of x^2, and is denoted by $f(x^2)$.

To denote a function containing two independent variables, as x, y, we inclose the variables and place the sign of function before them; thus $ay+bx^2$ may be expressed *generally* by $f(x, y)$.

(2.) The same functional letter is employed to denote the same *form* of expression when the variable and constant quantities are expressed by different letters. Thus, if $\mathrm{F}x$ denote $ax+x^2$, then $\mathrm{F}y$ will denote $by+y^2$. Also, if $\mathrm{F}x = ax+x^3$, $\mathrm{F}\phi x = a\phi x+(\phi x)^3$. For ex. if $\phi x = \sqrt{x}$, then $\mathrm{F}\phi x = a\sqrt{x}+(\sqrt{x})^3$.

(3.) If fx denote a function of x, then $f^{-1}x$ denotes the *inverse* function of x. Thus, if $fx =$ log. x, then $f^{-1}x =$ the *number* of which x is the logarithm. If $fx = \sin x$, then $f^{-1}x$ denotes the *arc* of which x is the sine.

EXERCISES.

1. If $fx = \sqrt{ax^3+b^2x}$ and $\phi x = (a+x)^2$, what is the value of $f\phi x$?

2. If $fx = \sqrt{x}$, what is the numerical value of $f^{-1}x$ when $x = 16$?

(4.) If an equation be *necessarily* true for every value of the variable which it contains, it is called a *Functional Equation.**

Thus, if $fx = ax$, then $f(2x) = a(2x) = 2ax = 2 \times fx$.
Also, if $fx = x^n$, $fy = y^n$, then $fx \times fy = x^n \times y^n -$

$$(xy)^n = f(xy).$$

EXERCISES.

1. If $fx = a^x$, and $fy = a^y$, it is required to show that $fx \times fy = f(x+y)$?

2. If $fx = ax + bx$, show that $fx \times fy = f(x+y)$.

(5.) The learner must have observed, that when $x+h$ is substituted for x in all the functions which he has yet developed, the *first term* of the developement is always the same as the primitive function, and the other terms succeed each other according to the ascending powers of h.

Let the function be x^3, then $(x+h)^3 = x^3 + 3x^2h + 3xh^2 + h^3$, in which the first term is x^3, and the others contain h, h^2, h^3, in succession.

If we employ the notation of functions this property will be expressed thus:

$$f(x+h) = fx + Ah + Bh^2 + Cx^3 + \&c.$$

The steps by which this is demonstrated are the following.

1. Let $h = 0$, then $fx = fx$, so that fx must be the first term of the developement.

* See De Morgan's Algebra, page 203.

2. None of the powers of h can be fractional, or in the form $\sqrt{h}$, $\sqrt[3]{h}$, &c.

3. None of the exponents can be negative, for then $h^{1} = \frac{1}{h}$, and when $h = 0$, $\frac{1}{h} - \frac{1}{0} =$ infinity, which, on this supposition, would be equal to a finite quantity.*

It was on this property, viewed in all its generality, as applying to all functions, that La Grange founded his "Theory of Functions," or his mode of establishing the principles of the differential calculus without employing the ideas of motion or limits.

SECTION VI.

ON SUCCESSIVE DIFFERENTIATION AND INTEGRATION.

(1.) As the differentials of all expressions which contain x raised to any power also contain x raised to the next inferior power, and as the differential of the independent variable always retains the *same* value, we may consider the differential of a function as a *new function* and determine its differential accordingly.

Let $u - x^3$, then $du = 3x^2dx$. Now, since $3x^2dx$ contains x, we may differentiate it as a new function. Hence, since dx is *constant*, the differential of $3x^2dx$ is $6xdx \times dx$, or $6xdx^2$, which is called the *second* differential of u. Again, the differential of $6xdx^2$ is

* La Grange, Théorie des Fonctions Analytiques.

$6xdx^2 \times dx$, that is, $6dx^3$, which is the third differential of u. As this does not contain x, but only dx, the operation can be carried no farther. The second differential of u is written d^2u, the third d^3u, &c. Or, according to Newton's notation, by $\ddot{u}$, $\dddot{u}$, &c.

$$\text{Hence } du = 3x^2dx, \text{ and } \frac{du}{dx} = 3x^2,$$

$$d^2u = 6xdx^2, \text{ and } \frac{d^2u}{dx^2} = 6x.$$

$$d^3u = 6dx^3, \text{ and } \frac{d^3u}{dx^3} = 6.$$

The learner must not confound d^2u with du^2, the former denoting the *differential of the differential of u*, and the latter the *square of the differential of u*.

EXERCISES.

1. Determine the successive differentials of ax^4?
2. Required the successive differentials of $(a+x^2)^3$?

(2.) Since a function admits of a series of successive differentials, a differential may have a successive number of integrals. Thus, if we have given the third differential of a function, we may obtain the primitive function after three successive integrations.

Ex. Required the function of which $6dx^3$ is the third differential?

$$\int 6dx^3 = 6\int dx^3 - 6xdx^2.$$

Again, $\int 6xdx^2 = 6\int xdx^2 = 3x^2dx$.

Lastly, $\int 3x^2dx = 3\int x^2dx = x^3$, the function required.

This process is denoted by placing the sign of integration before the differential as many times as there are to be integrations, or by placing a small figure above the sign, denoting the number of times.

$$\text{Thus, } \int\int\int 6dx^3, \text{ or } \int^3 6dx^3 = x^3.$$

Ex. Required the function of which $120a^2x^2dx^3$ is the *third* differential?

MACLAURIN'S THEOREM.

(3.) It is often required to develope an algebraic expression into an infinite series, and determine the particular law of the series. The first example of this kind which occurs to the pupil in his mathematical studies is generally the expansion of $\frac{1}{9}$ into an infinite series of decimal fractions. He has also had many examples of this general problem in the division of algebraic fractions, the extraction of the square or cube roots, the binomial theorem, and the theory of indeterminate coefficients. The theorem which we are now to investigate will put him in possession of another method of effecting the same thing by successive differentiations.

Let u represent any function of x, as, for example, $(a+x)^n$, $\sqrt{a+x^2}$, and let us suppose that this function when expanded will contain the ascending powers of x, and coefficients not containing x, which are to be determined. Let these coefficients be represented by A, B, C, &c. then $u = \text{A} + \text{B}x + \text{C}x^2 + \text{D}x^3 + \text{E}x^4 +$ &c. (1)

If we differentiate this equation and divide both sides

by dx, we have $\frac{du}{dx} = \text{B} + 2\text{C}x + 3\text{D}x^2$ &c. Differentiating and dividing by dx, and continuing the process, it is obvious the coefficients A, B, C, &c. will disappear in succession, and the result will be as follows:

$$\frac{du}{dx} = \text{B} + 2\text{C}x + 3\text{D}^2 + 4\text{E}^3 + \text{ \&c.}$$

$$\frac{d^2u}{dx^2} = \quad + 2\text{C} \ + 2.\,3\text{D}x \ + 3.\ 4\text{E}x^2 + \text{ \&c.}$$

$$\frac{d^3u}{dx^3} = \qquad + 2.\,3\text{D} \quad + 2.\,3.\,4\text{E}x + \text{ \&c.}$$

Let (u) denote the *limit* to which the given function approaches as x diminishes, and ultimately becomes 0. And $\left(\frac{du}{dx}\right)$ the limit of the differential coefficient when $x = 0$;

$\left(\frac{d^2u}{dx^2}\right)$ the limit of the value of $\frac{d^2u}{dx^2}$ when $x = 0$, &c.

Then

$$(u) = \text{A};\ \left(\frac{du}{dx}\right) = \text{B};\ \left(\frac{d^2u}{dx^2}\right) = 2\text{C};\ \left(\frac{d^3u}{dx^3}\right) = 2.3\text{D}, \text{ \&c.}$$

And $\text{A} = (u)$; $\text{B} = \frac{du}{dx}$; $\text{C} = \frac{1}{2}\left(\frac{d^2u}{dx^2}\right)$;

$$\text{D} = \frac{1}{2.\,3}\left(\frac{d^3u}{dx^3}\right);\ \text{E} = \frac{1}{2\,.\,3\,.\,4}\left(\frac{d^4u}{dx^4}\right)x^4;\ \text{\&c.}$$

Substituting these values for A, B, C, &c. in eq. (1)

$$u = (u) + \left(\frac{du}{dx}\right)x + \frac{1}{2}\left(\frac{d^2u}{dx^2}\right)x^2 + \frac{1}{2.\,3}\left(\frac{d^3u}{dx^3}\right)x^3 + \text{\&c.}$$

which is Maclaurin's Theorem.

EXAMPLES.

1. Expand $\frac{1}{1-x}$ into an infinite series.

Let $u = \frac{1}{1-x}$, then $du = \frac{dx}{(1-x)^2}$; $\frac{du}{dx} = \frac{1}{(1-x)^2}$.

Also $\frac{d^2u}{d} = \frac{1}{(1-x)^3}$, by differ. and dividing by dx.

And $\frac{d^3u}{dx^3} = \frac{1}{(1-x)^4}$, by differentiating the last equation and dividing by dx, &c.

When $x = 0$, $(u) = 1$; $\left(\frac{du}{dx}\right) = 1$; $\left(\frac{d^2u}{dx^2}\right) = 1$; &c.

Substituting 1 for these expressions in Maclaurin's Theorem we obtain $\frac{1}{1-x} = 1 + x + x^2 + x^3 + x^4 +$ &c.

2. Required the developement of $\sqrt{1+x}$?

Let $u = (1+x)^{\frac{1}{2}}$; then

$$du = \tfrac{1}{2}\,\frac{dx}{(1+x)^{\frac{1}{2}}} \text{ and } \frac{du}{dx} = \tfrac{1}{2}\,\frac{1}{(1+x)^{\frac{1}{2}}}.$$

Again $\frac{d^2u}{dx^2} = -\tfrac{1}{2}\cdot\tfrac{1}{2}\,(1+x)^{-\frac{3}{2}} = \frac{-\frac{1}{2}\cdot\frac{1}{2}}{(1+x)^{\frac{3}{2}}}$ by differentiating and dividing by dx.

Also $\frac{d^3u}{dx^3} = \frac{\frac{1}{2}\cdot\frac{1}{2}\cdot\frac{3}{2}}{(1+x)^{\frac{5}{2}}}$, &c. &c. Hence

$(u)=1$; $\left(\frac{du}{dx}\right) = \tfrac{1}{2}$; $\left(\frac{d^2u}{dx^2}\right) = -\tfrac{1}{2}\cdot\tfrac{1}{2}$; $\left(\frac{d^3u}{dx^3}\right) = \tfrac{1}{2}\cdot\tfrac{1}{2}\cdot\tfrac{3}{2}$.

Substituting these values in Maclaurin's Theorem,

$$\sqrt{1+x} = 1 + \tfrac{1}{2}x - \tfrac{1}{2}\cdot\tfrac{1}{2}\,x^2 + \tfrac{1}{2}\cdot\tfrac{1}{2}\cdot\tfrac{3}{2}\,x^3 - \text{\&c.}$$

EXERCISES.

1. Required the developement of $\sqrt{a+x}$ by Maclaurin's Theorem?

2. Expand $(a+x)^n$ by Maclaurin's Theorem, and also by the Binomial Theorem.

3. Required the series which expresses the cube root of $a+bx$?

4. Required the developement of $\frac{1}{a+x^2}$?

5. Expand $\frac{1}{\sqrt{a+x^2}}$ into an infinite series.

6. Required the series which expresses the cube root of 9, by making $9 = 1+8$?

TAYLOR'S THEOREM.

(4.) The theorem which we are now to investigate was first given by Dr. Brook Taylor, and is of great use in expanding functions into infinite series. The following is the principle on which it is founded. If in any function of x, x be changed into $x+h$, and if the differential coefficient of this new function be determined, that coefficient will be the same whether we suppose x to vary uniformly and h to remain constant, or h to vary and x to remain constant.

Let the function be $u = x^3$, then the new function is $u' = (x+h)^3$, the differential coefficient of which is $\frac{du'}{dx} = 3(x+h)^2$ on the supposition that x varies and h remains constant. Again, if h vary and x remain constant, the differential coefficient is $\frac{du'}{dh} = 3(x+h)^2$; hence $\frac{du'}{dx} = \frac{du'}{dh}$.

The principle is so evident as to be almost an axiom, for $x+h$ will produce the same change in the function whether x receive an increment k and h remain constant, or h receive the increment and x remain constant. For in both cases the new value of $x+h$ will be $x+h+k$.

(5.) Let u represent any function of x,
And u' the new function when x becomes $x+h$, and let $u' = u + \text{A}h + \text{B}h^2 + \text{C}h^3 + \text{\&c.}$ . . . . . (1)
in which A, B, C, &c. are unknown functions of x. Let x be supposed constant and h variable, then, by differentiating and dividing by dh, we have

$$\frac{du'}{dh} = \text{A} + 2\text{B}h + 3\text{C}h^2 + \text{\&c.} \quad . \; . \; . \; . \; . \; . \quad (2).$$

Again, let h be constant and x variable, then $du' = du + d\text{A}h + d\text{B}h^2 + \text{\&c.}$ Dividing by dx we have

$$\frac{du'}{dx} = \frac{du}{dx} + \frac{d\text{A}}{dx}h + \frac{d\text{B}}{dx}h^2 + \frac{d\text{C}}{dx}h^3 + \text{\&c.} \quad . \; . \; . \; . \quad (3)$$

Now, since the first sides of equations (2), (3) are equal, the two series on the second side are identical, and, by the theory of indeterminate coefficients, the coefficients of the same powers of h are equal.

Hence $\text{A} = \frac{du}{dx}$, $\text{B} = \frac{d\text{A}}{2dx}$, $\text{C} = \frac{d\text{B}}{3dx}$, $\text{D} = \frac{d\text{C}}{4dx}$, &c.

Substituting this value of A in the next equation, we have $\text{B} = \frac{1}{1.2}\frac{d^2u}{dx^2}$; substituting this value of B in the next, $C = \frac{1}{1.\,2.\,3}\,\frac{d^3u}{dx^3}$, &c. Introducing these values of A, B, C, &c. into equation (1) we have

$$u' = u + \frac{du}{dx}h + \frac{d^2u}{dx^2}\,\frac{h^2}{1.2} + \frac{d^3u}{dx^3}\,\frac{h^3}{1.\,2.\,3} + \text{\&c.}$$ which is Taylor's Theorem.

EXAMPLE.

Required the developement of $\sqrt{x+h}$ or $(x+h)^{\frac{1}{2}}$?

Let $u = \sqrt{x}$ and $u' = (x+h)^{\frac{1}{2}}$, then

$$du = \frac{dx}{2\sqrt{x}} \text{ and } \frac{du}{dx} = \frac{1}{2\sqrt{x}} = \tfrac{1}{2}x^{\frac{1}{2}}$$

Also $d^2u = -\frac{1}{2}\cdot\frac{1}{2}\,x^{-\frac{3}{2}}dx^2$, and $\frac{d^2u}{dx^2} - \frac{-\frac{1}{2}\cdot\frac{1}{2}}{x^{\frac{3}{2}}}$, &c.

Substituting these values in Taylor's Theorem, we have

$$\sqrt{x+h} = x^{\frac{1}{2}} + \tfrac{1}{2}\,x^{-\frac{1}{2}}h - \tfrac{1}{2}\cdot\tfrac{1}{2}\,x^{-\frac{3}{2}}\frac{h^2}{1.2} + \tfrac{1}{2}\cdot\tfrac{1}{2}\,\tfrac{3}{2}x^{-\frac{5}{2}}\frac{h^3}{1.2.3} - \text{\&c.}$$

That is, $\sqrt{x+h} = x^{\frac{1}{2}} + \frac{1}{2}\,\frac{h}{x^{\frac{1}{2}}} - \frac{1}{8}\cdot\frac{h^2}{x^{\frac{3}{2}}} + \frac{1}{16}\,\frac{h^3}{x^{\frac{5}{2}}} - $ &c.

EXERCISES.

1. Required the cube root of $x+h$?
2. Required the developement of the square root of 2?
3. Required the developement of $(a+x)^{\frac{1}{n}}$?

SECTION VII.

APPLICATION OF THE PRECEDING PRINCIPLES TO THE CALCULATION OF LOGARITHMS, AND OF SINES, TANGENTS, AND SECANTS.

(1.) To find the developement of log. $(1+x)$.

Let $u = \log.(1+x)$; then $du = \frac{d.(1+x)}{1+x} = \frac{dx}{1+x}$.

But $\frac{dx}{1+x} = dx - xdx + x^2dx - x^3dx +$ &c. by div.

Hence $u = \int dx - \int xdx + \int x^2dx - \int x^3dx +$ &c. by integ.

That is, $\log.(1+x) = x - \frac{x^2}{2} + \frac{x^3}{3} - \frac{x^4}{4} +$ &c. . . (1)

which is the series for the Naperian logarithm of $1+x$.

To find the developement of log. $(a+x)$ by *Maclaurin's Theorem.*

Put $u = \log.(a+x)$, then $(u) = a$, when $x = 0$,

$$\frac{du}{dx} = \frac{1}{a+x}, \text{ therefore } \left(\frac{du}{dx}\right) = \frac{1}{a}$$

$$\frac{d^2u}{dx^2} = -\frac{1}{(a+x)^2}, \text{ therefore } \left(\frac{d^2u}{dx^2}\right) = -\frac{1}{a^2}$$

$$\frac{d^3u}{dx^3} = \frac{2}{(a+x)^3}, \text{ therefore } \left(\frac{d^3u}{dx^3}\right) - \frac{2}{a^3}$$

$$\frac{d^4u}{dx^4} = -\frac{2 \cdot 3}{(a+x)^4}, \text{ therefore } \left(\frac{d^4u}{dx^4}\right) = -\frac{2 \cdot 3}{a^4}$$

&c. &c.

Substituting these values in Maclaurin's Theorem,

$$\log.\ (a+x) = \log.\ a + \frac{x}{a} - \frac{x^2}{2a^3} + \frac{x^3}{3a^4} - \&c.$$

Hence, when $a = 1$,

$$\log.\ (1+x) = x - \frac{x^2}{2} + \frac{x^3}{3} - \frac{x^4}{4} + \&c. \text{ as before.}$$

To find the developement of log. $(x+h)$ by *Taylor's Theorem.*

Let $u = \log.\ x$, and $u' = \log.\ (x+h)$, then

$$du = \frac{dx}{x}, \text{ and } \frac{du}{dx} = \frac{1}{x},$$

$$\frac{d^2u}{dx^2} = -\frac{1}{x^2};\ \frac{d^3u}{dx^3} = \frac{2}{x^3};\ \frac{d^4u}{dx^4} = -\frac{3}{x^4};\ \&c.$$

Substituting these values in Taylor's Theorem,

$$\log.\ (x+h) = \log.\ x + \frac{h}{x} - \frac{h^2}{2x^2} + \frac{h^3}{3x^3} - \frac{h^4}{4x^4} + \&c.$$

Or substituting a for x and x for h, we have

$$\log.\ (a+x) = \log.\ a + \frac{x}{a} - \frac{x^2}{2a^2} + \frac{x^3}{3a^3} - \frac{x^4}{4a^4} + \&c.$$

as before.

(2.) The successive terms of the series thus obtained for log. $(1+x)$ *diminish* in value when x is less than 1, and *increase* when x is greater than 1; but even in the former case the series *converges* so slowly as to be of no use for determining the numerical value of the logarithms of numbers.

By means of this series, however, we may deduce another, which will converge more rapidly, and is therefore better adapted for numerical calculation.

Substituting $-x$ instead of $+x$ in the series, we have

$$\log. (1-x) = -x - \frac{x^2}{2} - \frac{x^3}{3} - \frac{x^4}{4} - \&c.$$

Subtracting this series from series (1),

$$\log. (1+x) - \log. (1-x) = 2x + \frac{2x^3}{3} + \frac{2x^5}{5} + \frac{2x^7}{7} + \&c.$$

That is, $\log. \frac{1+x}{1-x} = 2x + \frac{2x^3}{3} + \frac{2x^5}{5} + \frac{2x^7}{7} + \&c.$

Let $\frac{1+x}{1-x} = 1 + \frac{z}{n}$, and therefore $x = \frac{z}{2n+z}$.

$$\text{And } \frac{1+x}{1-x} = \frac{1+\frac{z}{2n+z}}{1-\frac{z}{2n+z}} = \frac{2n+2z}{2n} = \frac{n+z}{n}.$$

Hence $\log. (n+z) - \log. n =$

$$\frac{2z}{2n+z} + \tfrac{2}{3}\left(\frac{z}{2n+z}\right)^3 + \tfrac{2}{5}\left(\frac{z}{2n+z}\right)^5 + \&c.$$

This series converges with sufficient rapidity to determine the logarithm of $n+z$, provided we know the logarithm of n.

EXAMPLE.

Required the logarithm of 2?

Let $n = 1$ and $z = 1$, then, since $\log. 1 = 0$, we have $\log. 2 = \frac{2}{3} + \frac{2}{3} \times \frac{1}{3^3} + \frac{2}{5} \times \frac{1}{3^5} + \&c.$

The learner will now find the value of *seven* terms of this series in decimal fractions, the sum of which will give him the number .6931471, which is the *Naperian* logarithm of 2, and this number multiplied by .43429448 will give .322219, the *common* logarithm of 2.

EXERCISES.

1. Let $n = 2$, $z - 1$, then log. $(2+1) =$

$$\log. 2 + 2 \left\{ \frac{1}{5} + \frac{1}{3 \cdot 5^3} + \frac{1}{5 \cdot 5^5} + \frac{1}{7 \cdot 5^7} + \&c. \right\}$$

from which the learner is required to find the log. of 3?

2. Required the log. of 4?

3. Required the log. of 5, 6, 7, 8, 9?

NOTE. The converse of this problem, viz. "To find the number corresponding to a *given* logarithm," has already been investigated under the name of the "Exponential Theorem."

(3.) To find the developement of sin x.

1. By Maclaurin's Theorem.

Let $u = \sin x$, then $(u) = 0$, when $x = 0$.

Again, $\dfrac{du}{dx} - \cos x$, and $\left(\dfrac{du}{dx}\right) = 1$.

$$\frac{d^2u}{dx^2} = -\sin x, \;..\; \left(\frac{d^2u}{dx^2}\right) = 0.$$

$$\frac{d^3u}{dx^3} = \sin x, \;..\; \left(\frac{d^3u}{dx^3}\right) = -1.$$

&c. &c.

Substituting these values in the theorem,

$$\sin x = x - \frac{x^3}{1.2.3} + \frac{x^5}{1.2.3.4.5} - \frac{x^7}{1.2.3.4.5.6.7} + \&c.$$

2. By Taylor's Theorem.

Let $u = \sin x$, and $u' = \sin (x+h)$, then

$$\frac{du}{dx} = \cos x; \; \frac{d^2u}{dx^2} = -\sin x; \; \frac{d^3u}{dx^3} = \cos x, \&c.$$

Substituting these values of $\frac{du}{dx}$, $\frac{d^2u}{dx^2}$, &c. in Taylor's Theorem,

$$\sin(x+h) = \sin x + \cos xh - \sin x \frac{h^2}{1.2} - \cos x \frac{h^3}{1.2.3} + \text{\&c.}$$

Let $x = 0$, then since $\cos x =$ radius $= 1$, the equation becomes $\sin h = h - \frac{h^3}{1.2.3} + \frac{x^5}{1.2.3.4.5}$, which is the same as before.

Obs. The learner may calculate the natural sine of 1, 2, 3, &c. degrees, by taking the radius $=100000$, and compare his results with those contained in a table of natural sines. Having calculated the sines, he can easily obtain the cosines, tangents, &c.

EXERCISES.

1. Required the log. of .1 ?
2. Required the log. of .02 ?
3. Required the developement of $\cos x$, both by Maclaurin's and Taylor's Theorem ?
4. Required the developement of $\tan x$?

OBSERVATIONS ON THE PRECEDING PART.

1. We have endeavoured to simplify this Part as much as possible by introducing numerical and geometrical illustrations. The necessity of adopting this measure as often as possible cannot be too forcibly impressed on the teacher, as without this practical application, the learner frequently thinks he understands the subject, when, in fact, his ideas are exceedingly obscure and ill-defined. This is particularly the case with regard to the differentials of exponential and logarithmic functions.

2. The mode which has been adopted for finding the differentials of circular functions requires a much less acquaintance with trigonometry than that usually employed. The pupil will therefore be able to read this Part without having an extensive knowledge of trigonometrical formulæ, or the "Arithmetic of Sines."

PROMISCUOUS QUESTIONS AND EXERCISES.

1. Explain the principle called the *Theory of Indeterminate Coefficients,* or rather of *coefficients to be determined.*

2. How would you find the square root of 2 by logarithms?

3. How would you find the square root of 2 by Maclaurin's Theorem?

4. How would you find the square root of $\frac{2}{3}$ by logarithms?

5. If the base of the system of logarithms be 10, how would you find the differential of a given number, suppose 20?

6. How would you find the differential of a given number, suppose 20, if the base of the system be 2.71828?

7. What distinction do you make between *algebraic* and *transcendental* functions?

8. Give the names of the following functions of x, $(\sin x)^2$; $(ax+bx^2)^{\frac{1}{n}}$; a^x; $\log. (a+x)$.

9. Can you express in finite numbers the integral of $\frac{dx}{\sqrt{1-x^2}}$?

10. Describe and illustrate by examples what you understand by *inverse functions*.

11. What is meant by successive differentiations?

12. State the principle on which Maclaurin's Theorem is founded, and repeat the theorem.

13. Do the same for Taylor's Theorem.

14. Can the logarithm of a *prime* number be expressed by a *finite* number?

15. What is the principle on which La Grange's theory of functions is founded?

16. When a theorem is said to *fail* when applied to a *particular* question, does the theorem itself really fail, or do you *fail* in not perceiving that you are applying a rule to solve a question which does not come under that rule?

17. What is the differential of $\log. \sqrt{x}$?

18. What is the differential of $\log. \frac{x}{y}$, and what is its numerical value, if x increase uniformly at the rate of 1, and y at the rate of 2, when x becomes 3 at the same moment that y becomes 4?

19. What is the differential of $\log. (a+x)^3$?

20. What is the value of $\int_c^e 2(a+x)\,dx$, when $c=10$, $e=20$, and $a=4$?

21. Illustrate the exercise by a geometrical figure.

22. What is the value of $\int_a^b 3(c+nx^2)^2 2nx\,dx$ when $a=4$, $b=6$, $c=4$, and $n=2$?

23. Illustrate this example geometrically.

24. What is the value of $\int_a^b \frac{dx}{c+x}$, when $a=2$, $b=3$, and $c=4$?

25. What is the value of $\int_c^e \frac{dx}{\sqrt{a^2-b^2x^2}}$, when $a=200000$, $b=2$, $c=50000$, and $e=60000$?

26. Illustrate the preceding example geometrically.

PART IV.

APPLICATION OF THE PRECEDING PRINCIPLES TO DETERMINE THE RADIUS OF CURVATURE, NATURE OF EVOLUTES, &c. OF THE MORE USEFUL CURVES OF THE SECOND AND HIGHER ORDERS.

In the practical applications of the elementary principles contained in the First Part, we confined ourselves to the properties of the circle and the conic sections. In this part we shall introduce the pupil to the knowledge of two of the curves of the higher orders which are of the greatest use in the study of certain parts of Natural Philosophy.

SECTION I.

RADIUS OF CURVATURE—INVOLUTES AND EVOLUTES.

(1.) If a circle touch a curve in a given point, and if the radius be of such a length that a greater circle cannot be described through that point without cutting the curve, it is called the *Circle of Curvature*, and its radius, the *Radius of Curvature.* This circle is sometimes called the *Equicurve Circle*, and frequently the *Osculating Circle.*

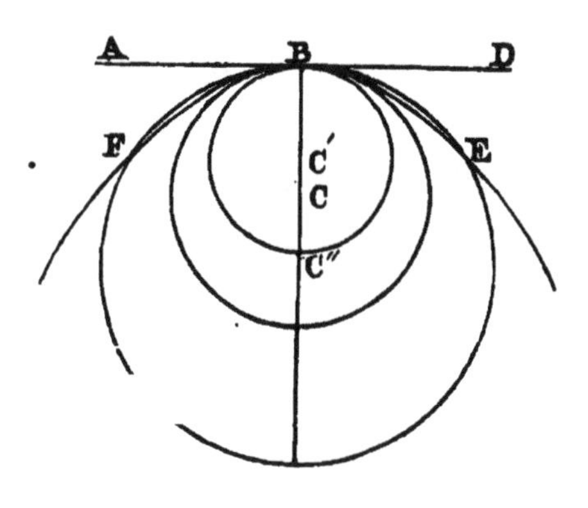

Let FBE be a curve, and ABD a straight line touching it in B. Draw BC at right angles to AD, and in BC produced take a point, so that a circle described about that point as a centre with the radius C″B may cut the curve in two points F, E. Let the radius of the circle be taken less and less, then it is obvious that the points F, E will approach the point B, and ultimately coincide with it. In this state the circle is that of equal curvature.

Since the circle and curve have the same degree of bending or curvature at the point B, the circle of curvature may be viewed as that which is equally deflected from the common tangent, or as the limit of a circle passing through three points in the curve, when the two extreme points continually approach the intermediate point, and ultimately coincide with it.

(2.) In every curve, except the circle, the radius of curvature is a variable quantity, and the centres of the circles of curvature, at different points of the curve, have different positions. If the centres of curvature be therefore determined for an indefinitely great number of points, and a line drawn through these points, that line will be a curve, the nature of which will depend on that of the given curve. Hence if a *mould* be formed, having that curvature, the given curve may be described by continued motion.

Let the curve, for example, be such, that its centres of curvature lie in the arc of a circle of a given diameter. Take a circle of that diameter, suppose a halfpenny, and fix a thread having a loop at one of its extremities to the circumference of the circle by means of cement or sealing-wax. Roll the thread about the circumference, place it on the surface of the paper, and, holding the point of a pencil in the loop, *unwind* the thread from the circle, keeping it always tight, and the pencil will trace out the curve required. The curve $Acc'c''$, from which the curve $Aaa'a''$ is evolved, is called the *Evolute*, and the curve thus traced, its *Involute*.

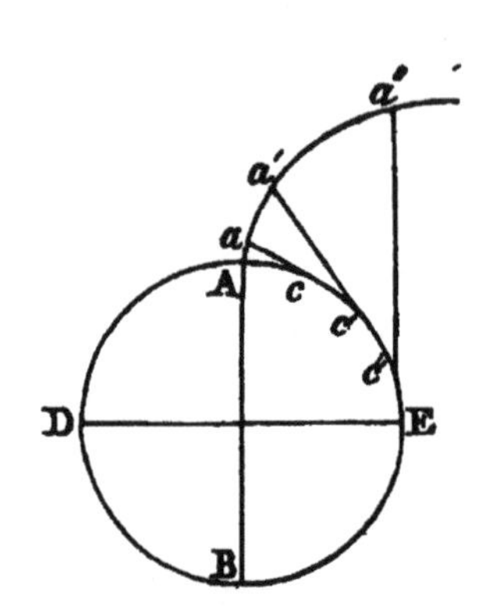

Several obvious consequences will present themselves to the pupil.

1. The points c, c', c'', &c. are the centres of curvature of the points a, a', a'', &c. and ca, $c'a'$, &c. the radii of curvature of the arcs at those points.

2. The lines ca, $c'a'$, &c. are tangents to the evolute.

3. The radius of curvature is at right angles to the tangent of the involute at the given point.

4. The radius of curvature is equal to the arc of the evolute, reckoned from the point A, where the curve

commences. Thus, the arc AC $\doteq$ ca, the radius of curvature at the point a.

(3.) To determine the radius of curvature at a given point in a given curve.

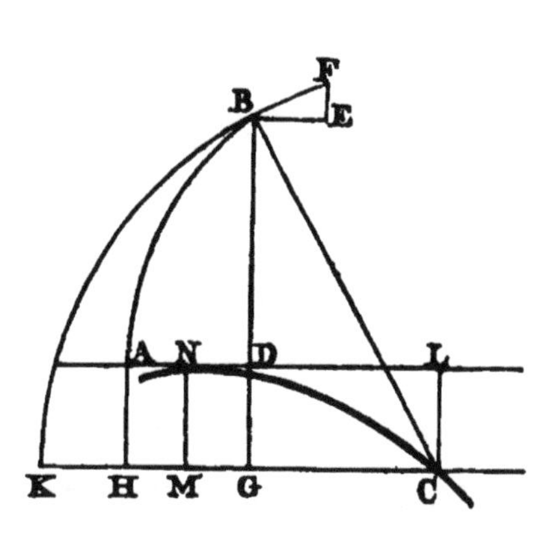

Let AB be the arc of a given curve, of which NC is the evolute, and let BF be a tangent at B, and BC a line at right angles to BF; then BC is the radius of curvature. Draw the axis AD and the other lines as in the figure. From C, with the radius CB, describe an arc of a circle, and draw CK parallel to AD.

Let KH $= a$, GD $= b$, AD $= x$, DB $= y$, and BC $= r$. Then KG $= a + x$, and BG $= b + y$. Also, BE$=dx$, EF $= dy$, and BF $\doteq dz$, the arc AB being denoted by z. Then, by the equation of the circle,

$$(b + y)^2 = (2r - a - x)(a + x), \text{ that is}$$

$$b^2 + 2by + y^2 = 2ra + 2rx - 2ax - x^2 - a^2,$$

Differentiating this equation and dividing by 2,

$$bdy + ydy = rdx - adx - xdx. \text{ Differ. this equation}$$

$$bd^2y + yd^2y + dy^2 = rd^2x - ad^2x - xd^2x - dx^2, \text{ or}$$

$$dy^2 + dx^2 = (r - a - x)d^2x - (b + y)d^2y \ldots (1)$$

But since the given curve and the circle of curvature coincide at the point B and have the same tangent, the differentials of the abscissæ, of the ordinates and of the arcs of these curves are equal.

Now, since the triangles BEF, BGC, are similar

$$dz : dy :: r : \text{GC}, \text{ Hence}$$

GC, or $r - \dot{a} - x = \frac{rdy}{dz}$. Again,

$dz : dx :: r :$ BG; hence BG, or $b + y = \frac{rdx}{dz}$.

Substituting these values, and dz^2 for its equal, $dy^2 + dx^2$, in equation ...... (1)

we have $z^2 = \frac{rdyd^2x}{dz} - \frac{rdxd^2y}{dz}$. Hence, by reduction

$$r = \frac{dz^3}{dyd^2x - dxd^2y} \text{.............(2)}$$

Cor. If we consider in succession each of the three variable quantities x, y, z, as the independent variable, the above general formula will be adapted to each of these cases. Since the differential of the independent variable is a *constant* quantity, which we may assume equal to 1, its *second* differential will be equal to 0.

1. When x is the independent variable,

$$= \frac{dz^3}{-dxd^2y} = \frac{dz^3}{-d^2y}, \text{ when } dx = 1 \text{ .... (1)}$$

2. When y is the independent variable,

$$r = \frac{dz^3}{dyd^2x} = \frac{dz^3}{d^2x}, \text{ when } dy = 1, \text{ .... (2)}$$

3. When z is the independent variable,

$$r = \frac{dzdy}{d^2x} = \frac{dy}{d^2x} \text{ when } dz = 1, \text{ .... (3)}$$

EXAMPLE 1.

Required the radius of curvature for a point in the parabola, whose abscissa is x, and ordinate y, the parameter being p.

Let x be the independent variable, and $dx = 1$.

Since $y^2 = px$, $y = \sqrt{px} = p^{\frac{1}{2}}x^{\frac{1}{2}}$,

$$dy = \tfrac{1}{2}p^{\frac{1}{2}}x^{-\frac{1}{2}}dx = \frac{p^{\frac{1}{2}}}{2x^{\frac{1}{2}}}, \text{ differ. this equation}$$

$$d^2y = -\tfrac{1}{4}p^{\frac{1}{2}}x^{-\frac{3}{2}} = -\frac{p^{\frac{1}{2}}}{4x^{\frac{3}{2}}}. \text{ But}$$

$$dz = \sqrt{dx^2 + dy^2} = \sqrt{1 + \frac{p}{4x}} = \tfrac{1}{2}\sqrt{\frac{4x+p}{p}}$$

Hence $\dfrac{dz^3}{-d^2y} = \dfrac{(4x+p)^{\frac{3}{2}}}{2\sqrt{p}} =$ radius of curvature required.

When $x = 0$, the expression for the radius of curvature becomes $\frac{1}{2}p$, which is the length of the radius of curvature at the vertex.

EXERCISE.

Required the length of the radius of curvature for a point in a parabola, whose abscissa is 9, and ordinate 6, the ordinate being considered as the independent variable?

EXAMPLE 2.

Required the radius of curvature at a point, in an ellipse, whose abscissa is x, and ordinate y, the major axis being a, and the minor b?

Let x be the independent variable, and $dx = 1$.

Since $y^2 = \dfrac{b^2}{a^2}(ax - x^2)$, we have

$2ydy = \dfrac{b^2}{a^2}(a - 2x)\,dx$. Differentiating this equation, and making $dx = 1$, we have

$$2yd^2y + 2dy^2 = \frac{b^2}{a^2} \times - 2dx \times dx = - \frac{2b^2}{a^2}$$

Hence $dy - \frac{b^2(a-2x)}{2a^2y}$ and $- d^2y = \frac{a^2dy^2 + b^2}{a^2y}$

Substituting the values of y and dy in these equations, they become

$$dy = \frac{b\,(a-2x)}{2a\sqrt{ax-x^2}} \text{ and} - d^2y -$$

$$\frac{(a-2x)^2a^2b^2}{4a^3c(ax - x^2)\sqrt{(ax-x^2)}} + \frac{b}{a\sqrt{ax-x^2}} = \frac{ab}{4(ax - x^2)^{\frac{3}{2}}}$$

By reduction.—Therefore

$$dz = \sqrt{dy^2 + dx^2} = \sqrt{\left\{\frac{b^2(a - 2x)^2}{4a^2(ax-x^2)} + 1\right\}}$$

$$= \frac{1}{2a}\sqrt{\left\{\frac{a^2b^2+(a^2-b^2)\,(4ax-4x^2)}{ax-x^2}\right\}}$$

Hence $r = \frac{dz^3}{-d^2y} = \frac{(a^2b^2)+(a^2-b^2)(4ax-4x^2)^{\frac{3}{2}}}{2a^4b}$, the radius required.

EXERCISES.

1. Required the radius of curvature at the extremity of the major axis of an ellipse, whose major axis is 10, and minor 6?

2. Required the radius of curvature at the extremity of the minor axis of the same ellipse?

(4.) Given the equation of the involute to determine the nature of the evolute.

By means of the preceding problem we determine the lengths of NL and LC (fig. p. 139) which are the co-ordinates of the point C in the evolute, referred to NL as the axis, and hence the nature of the curve will be determined.

1. Let x be the independent variable, and $dx=1$.

Then, since the triangles BEF, BGC, are similar,

$$\text{GC or DL} = \frac{rdy}{dz} = \frac{dz^3}{-d^2y} \times \frac{dy}{dz} = \frac{dydz^2}{-d^2y}.$$

$$\text{Hence AL} = x + \frac{dydz^2}{-d^2y} \ldots\ldots\ldots\ldots (1)$$

And $\text{NL}=\text{AL}-\text{AN}=$ the abscissa of the evolute. Again,

$$\text{DG or CL} = \text{BG}-\text{BD} = \frac{rdx}{dz} - y = \frac{dz^2}{-d^2y} - y \ldots (2)$$

2. Let y be the independent variable, and $dy=1$.

$$\text{Then AL} = x + \frac{dz^2}{d^2x} \ldots\ldots\ldots\ldots (3)$$

$$\text{And LC} = \frac{dxdz^2}{d^2z} - y \ldots\ldots\ldots\ldots (4)$$

3. Let z be the independent variable, and $dz=1$.

$$\text{Then AL} = x + \frac{dy^2}{d^2x} \ldots\ldots\ldots\ldots (5)$$

$$\text{And LC} = \frac{dxdy}{d^2x} - y \ldots\ldots\ldots\ldots (6)$$

EXAMPLE.

Required the nature of the curve whose *involute* is the common parabola?

Let x be the independent variable and $dx = 1$,

Then $y^2 = px$ and $y = p^{\frac{1}{2}}x^{\frac{1}{2}}$. Differentiating,

$dy = \frac{1}{2}p^{\frac{1}{2}}x^{-\frac{1}{2}}dx = \frac{p^{\frac{1}{2}}}{2x^{\frac{1}{2}}}$. Differentiating this equation, and omitting the multiplier dx,

$$d^2y = -\frac{1}{4}p^{\frac{1}{2}}x^{-\frac{3}{2}} = -\frac{p}{x^{\frac{3}{2}}}.$$

Also $dz^2 = dx^2 + dy^2 = 1 + \frac{p}{4x} = \frac{4x+p}{4x}$.

But DL $= \frac{dydz^2}{-d^2y} = 2x + \frac{1}{2}p$, by substitution.

Hence AL $= 3x + \frac{1}{2}p$. But the radius of curvature at the vertex was already found to be $\frac{1}{2}p$. Hence NL $= 3x =$ the abscissa of the evolute.

Again CL $= \frac{dz^2}{-d^2y} - y = \frac{4x^{\frac{3}{2}}}{\sqrt{p}}$, by substitution.

Let NL be denoted by X and CL by Y,

then $X = 3x$ and $x = \frac{X}{3}$. Hence $x^3 = \frac{X^3}{27}$;

and $Y = \frac{4x^{\frac{3}{2}}}{\sqrt{p}}$, and $x^{\frac{3}{2}} = \frac{Y\sqrt{p}}{4}$. Hence $x^3 = \frac{Y^2p}{16}$.

Therefore $\frac{X^3}{27} = \frac{pY^2}{16}$, and $X^3 = \frac{27p}{16}Y^2$.

Hence the evolute is a semicubical parabola, whose parameter is $\frac{27}{16}$ of the parameter of the involute.

NOTE. We shall give some of the more useful applications of the preceding problem in the following sections, particularly with regard to the curve called a *cycloid*.

SECTION II.

ON THE LOGARITHMIC CURVE.

(1.) In the straight line AD take any number of equidistant points, and draw perpendiculars at those points, and take these perpendiculars in geometrical progression, then the curve which passes through their extremities will be the *Logarithmic Curve.*

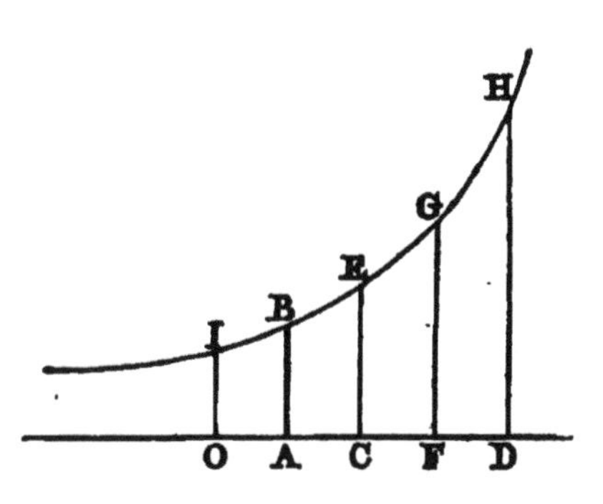

Let AB $= a$, the base, and OI $= 1$, then

since $1 : a :: a :$ CE, CE $= a^2$,

and $a : a^2 :: a^2 :$ FG, FG $= a^3$

Hence OA is the logarithm of AB; OC of CE; OF of FG; &c.

If the ratio of OI to AB be that of 1 to 10, the curve is that belonging to the common system. If the ratio be that of 1 to 2.30258509, the curve belongs to the Naperian system.

As logarithms may be continued *ad infinitum* below the base, it is obvious that the curve continually approaches the indefinite straight line DO, but never meets it.

The line OD is called the *axis;* any line, OD, reckoned on the axis from O, an *abscissa;* and DH the *ordinate.* Hence the equation of the curve is $y=a^x$.

(2.) To draw a tangent to the curve at the point B.

Let A denote the Naperian logarithm of a, and y' the Naperian logarithm of y; then

Since $y = a^x$, log. of y = log. of a^x;
that is, $y' = x\text{A}$. Differentiating this equation,

$$dy' = \text{A}dx\text{; but } dy' = \frac{dy}{y}. \text{ Therefore}$$

$$\text{A}dx = \frac{dy}{y}, \text{ and } \frac{ydx}{dy} = \frac{1}{\text{A}}.$$

But $\frac{ydx}{dy}$ is the expression for the subtangent, and $\frac{1}{\text{A}}$ is the modulus of the system.

Hence, if the curve belong to the common system, let fall the perpendicular BC, take CD = .43429448, and join DB, which will be the tangent required.

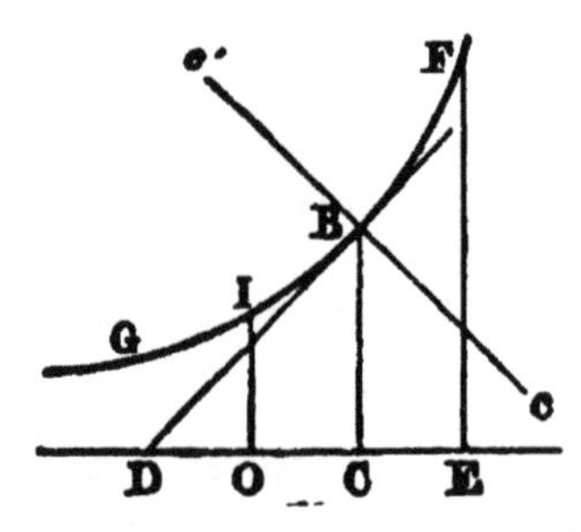

If the curve belong to the Naperian system, take CD = 1, and DB will be the tangent to the point B.

COR. The subtangent is the same for every point in the curve.

(3.) Let any two logarithmic curves pass through the point I, and let B, D be two points having equal ordinates, BC, DE, and let BF, DG be tangents at those points, Then OF : OE :: FC : GE.

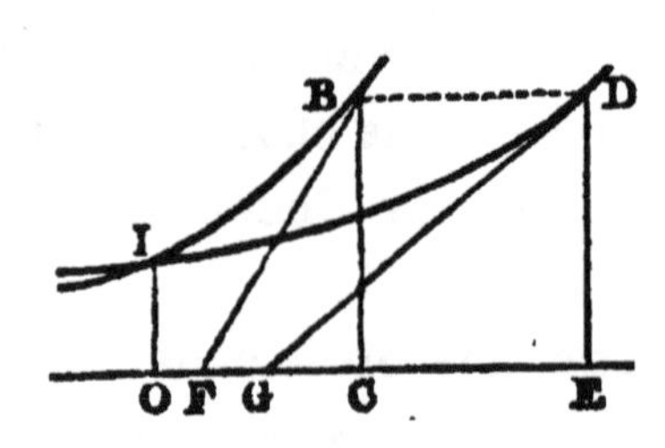

For OF, OB are the logarithms of the equal numbers represented by CB, ED, and FC is the modulus of one of the systems, and GE of the other; and since the logarithms of equal numbers are proportional to the moduli of the systems, the truth of the proposition is obvious.

(4.) Since OC is the logarithm of CE (fig. p. 145) and OD of DH, OD − OC = CD = log. of DH − log. of CE = log. $\frac{\text{DH}}{\text{CE}}$.

COR. If a, b, c, d, be any four ordinates, then the following proportion evidently follows from the preceding property:

$a-b$: $c-d$ :: log. $\frac{b}{a}$: log. $\frac{d}{c}$, whatever be the system to which the curve belongs.

(5.) Required the radius of curvature at a point B (fig. 1st, p. 146) whose abscissa is x, and ordinate y.

Let A denote the Naperian log. of a, and m the modulus of the system, then

$$\frac{1}{\text{A}} = m, \text{ and } \text{A} = \frac{1}{m},$$

$$\text{Since } y = a^x,\ dy = \text{A}a^x dx = \frac{ydx}{m}.$$

Let x be the independent variable, and $dx = 1$, then $dy - \frac{y}{m}$. Differentiating this equation,

$$d^2y = \frac{dy}{m}, \text{ and}$$

$$dz = \sqrt{dx^2 + dy^2} = \sqrt{1 + \frac{y^2}{m^2}} = \frac{(m^2 + y^2)^{\frac{1}{2}}}{m}.$$

Hence $r = \frac{dz^3}{-d^2y} = \frac{(m^2+y^2)^{\frac{3}{2}}}{-my}$, the radius of curvature required.

Since the divisor *my* has the sign *minus* before it, the value of *r* is *negative*, and must therefore be reckoned in the opposite direction. Hence, draw BC′ at right angles to BD, produce it to C′, and make BC′ equal to the value of *r* for a particular ordinate, and C′ will be the centre of curvature for the arc at the point B.

(6.) To find the area of the space contained between an ordinate EF, the axis continued indefinitely towards the left, and the arc of the curve FBG continued indefinitely (fig. 1st, p. 146).

Let CB $= a$, EF $= y$, and OE $= x$,

then the subtangent DC $= \frac{ydx}{dy} = m$, the modulus.

Hence $ydx = mdy$; but ydx is the differential of the area CEFB, the integral of which will be the area.

Therefore $\int mdy = my + \text{C} =$ area of CEFB.

Now, to find the value of the *constant* C, let y move up to BC, and ultimately coincide with it, then

$y = a$, and the area of CEFB becomes O; hence

$my + \text{C} = 0$, or $ma + \text{C} = 0$, and $\text{C} = -ma$.

Hence the area of CEFB $= my - ma =$ CD(EF$-$CB).

Let CB move on to an infinite distance towards the left; it will continually diminish, and may be made less than any *finite* magnitude. Hence the limit of the area when CB $= 0$ is CD $\times$ EF.

COR. Hence the areas of the spaces contained between given ordinates and the infinite branches of the curve and axis are to each other as these ordinates.

(7.) By means of this curve we may find any number of geometrical proportionals between two given lines.

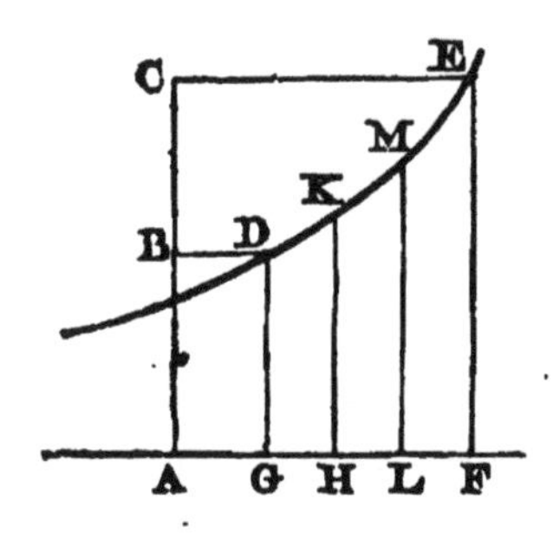

Let it be required to find two mean proportionals between two given lines AB, AC. Draw AC at right angles to the axis, and draw BD, CE parallel to the axis cutting the curve in the points D, E, and let fall the perpendiculars DG, EF on the axis. Divide GF into *three* equal parts, and draw HK, LM from the points of division at right angles to AF, and these lines will be the *two* mean proportionals required. For

$$GD = AB \text{ and } FE = AC.$$

And GD : HK :: HK : LM :: LM : FE by the nature of the curve.

COR. Since GD : FE :: $GD^3 : HK^3$, we may make FE the double of GD, and the proportion becomes,

1 : 2 :: $GD^3 : HK^3$. Hence, if GD be the side of a cube, HK will be the side of another cube, whose solidity is double that of the first. This is one of the solutions of the *famous* problem of *doubling the cube.*

OBS. It is said, that whilst the plague was raging in Attica, the *Delian* oracle was consulted in order to discover the means necessary to appease the wrath of the offended Deity. The answer returned was, that they must *Double the altar*, which was in the form of a cube. An altar was made, having twice the linear dimensions, yet the plague continued to rage. Plato and other geometers of eminence were applied to, but the geometry of the period was not sufficiently powerful for the solution of the problem. This circumstance is said to have led to the invention of some of the curves of the higher orders.*

EXERCISES.

1. Required the length of the radius of curvature at a point in the logarithmic curve belonging to the common system, whose abscissa OC is 4?

2. Required the area of the space infinitely long, contained between the ordinate corresponding to an abscissa OC of 4 inches, and the branch of the curve and continuation of the axis?

3. What is the area of the space CEFB, OC being equal to 3 and OE to 4, the curve belonging to the Naperian system?

* Montucla, Histoire des Mathématiques, tom. i. p. 174.

perhaps this is one way

To find contents of old cube doubled no of feet

& extracted ~~[illegible]~~ root & then made fresh cube

SECTION III.

ON THE CYCLOID.

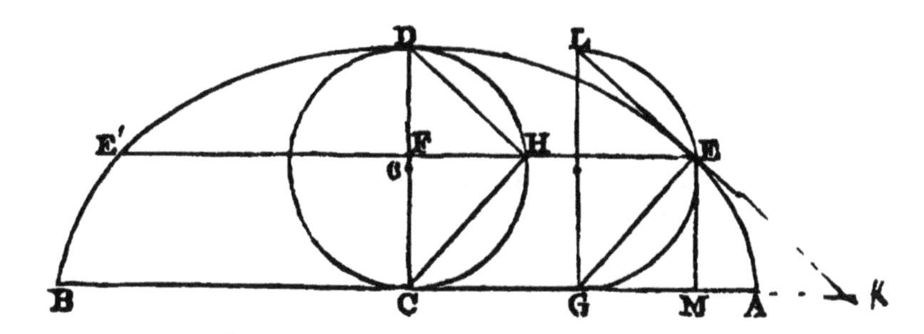

(1.) If a circle, whose diameter is CD, roll along the straight line AB from the point A, then the point in its circumference which coincided with A at the beginning of the motion, will trace out the curve AEDB, called a *Cycloid*. The circle which is supposed to roll along AB, is called the *generating circle*, and AB is obviously equal to the circumference of this circle. The straight line AB is called the *base* of the cycloid, and CD its *axis*. The point C is the *vertex*, EF the *ordinate* of the point E, and DE its *abscissa*.

The learner may construct this curve by fixing a small piece of black lead, by cement or sealing-wax, on the rim of a farthing or halfpenny, and rolling the coin along the edge of a straight rule laid flat on the surface of the paper, when the point will trace out the cycloid. The generating circle used for constructing the annexed figure was a sixpence.

(2.) Let EF be an ordinate at the point E, cutting the generating circle in H, then HE will be equal to the arc DH of the circle.

The arc EG=the line AG, and the semicircumference GEL=AC. Therefore the arc EL = the arc HD = CG =HE.

COR. Since FH = sin DH, the ordinate EF = arc DH + sin DH.

(3.) The tangent at E is parallel to the chord DH of the circle.

For the line EG may be viewed as turning round G in describing an indefinitely small portion of the curve, hence, from the property of the circle, the tangent at E is perpendicular to EG. But EG is parallel to CH and HD at right angles to CH, therefore the tangent EK is parallel to HD.

(4.) To draw a tangent to the curve at the point E.

Let the tangent EL be supposed produced to meet the axis CD produced, in the point K.

Let DF$=x$, EF$=y$, arc DH$=z$, DC$=2a$, and FH$=s$.

Then $y=z+s$, and $dy=dz+ds$.

But since s is the sine, and x the versed sine of z,

$$dz=\frac{adx}{s}, \text{ by equation (1) page 109.}$$

And $ds=\frac{1}{a}\cos z dz$ by equation (2) page 114. Substituting the value of dz from the preceding equation, and putting $a-x$ for $\cos z$, we have

$$ds=\frac{(a-x)\,dx}{s}. \text{ Hence the subtangent}$$

$$\text{FK}=\frac{ydx}{dy}=\frac{ydx}{dz+ds}=\frac{sy}{2a-x} \text{ by substitution.}$$

But since the triangles DFH, CFH, are similar,

$$\frac{DF}{FH} = \frac{FH}{FC} = \frac{}{2a - x}. \quad \text{Hence } FK = \frac{DF}{FH} \times FE.$$

(5.) To find the length of the arc DE of the cycloid.

Let z = arc DE, DC $= a$, DF $= x$, FE $= y$, and DH $= c$,

Then $s^2 = ax - x^2$ by the equation of the circle.

Hence $2sds = adx - 2xdx$, and $ds = \dfrac{adx - 2xdx}{2s}$

And $dz = \dfrac{adx}{2s}$, as in the preceding example.

But $y = z + s$, by the property of the cycloid.

Therefore $dy = dz + ds = \dfrac{(a-x)dx}{s}$ by substitution.

Consequently $dz - \sqrt{dx^2 + dy^2} =$

$$= dx\sqrt{\left(1 + \frac{(a-x)^2}{s^2}\right)} = \frac{dx}{s}\sqrt{a^2 - ax} =$$

$$= \frac{dx\sqrt{a^2 - x^2}}{\sqrt{ax - x^2}} = dx\sqrt{\frac{a}{x}} = a^{\frac{1}{2}}x^{-\frac{1}{2}}dx.$$

Hence $\int dz = \int a^{\frac{1}{2}}x^{-\frac{1}{2}}dx$, by integration,

That is $z = \dfrac{a^{\frac{1}{2}}x^{\frac{1}{2}}dx}{\frac{1}{2}dx} = 2\sqrt{ax} = 2$DH.

The arc DE is thus double the chord DH of the circle.

COR. 1. Hence the length of half the cycloidal curve is double the diameter of the generating circle, and the entire curve *four* times the length of the diameter.

COR. 2. Since the arc DE $= 2$DH, DE$^2 - 4$DH$^2 =$ 4DF $\times$ DC; therefore DF : DE :: DE : 4DC.

COR. 3. Since the arc AD$=2$DC and the arc ED$=2$DH

AD$^2=4$DC^2 and ED$^2=4$DH^2.

Therefore AD^2:ED^2::4DC^2:4DH^2: :DC^2:DH^2: :DC:DF.

That is, the abscissa DF varies directly as the square of the corresponding arc DE.

(6.) To find the radius of curvature at a given point in the cycloid, and the nature of its evolute.

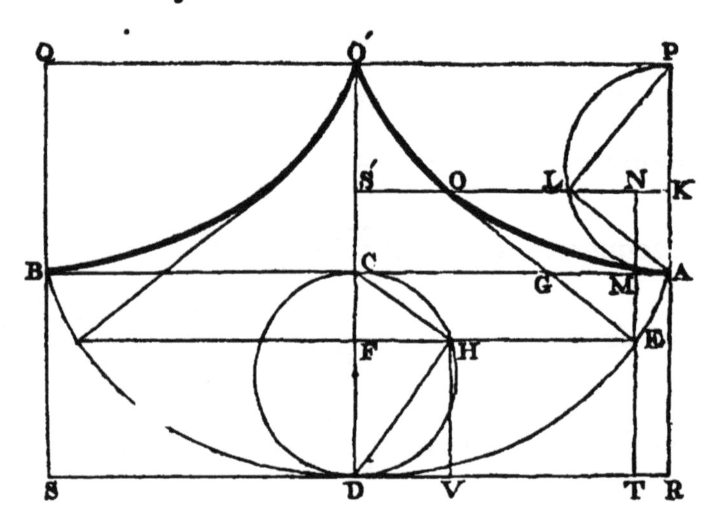

Let ADB be an inverted cycloid, and AOO′ the evolute of the arc AD. Let AD $= 2$DC $= a$, AE $= z$, AM $= x$, EM $= y$. Then DE $= a-z$, and DF $= \frac{1}{2}a-y$. Then, by the property of the curve contained in the preceding Cor.

$$a^2 : (a-z)^2 :: \tfrac{1}{2}a : \tfrac{1}{2}a-y; \text{ therefore}$$

$$y = \frac{2az-z^2}{2a}, \text{ and } dy = \frac{adz - zdz}{a} = \frac{a-z}{a},$$

when z is the independent variable and $dz = 1$.

Again, $dx^2 = dz^2 - dy^2 = 1 - \left(\frac{a-z}{a}\right)^2 = \frac{2az - z^2}{a^2}$;

therefore $dx = \frac{\sqrt{2az-z^2}}{a}$, and $d^2x = \frac{a-z}{a\sqrt{2az-z^2}}$.

Hence, the radius of cur. $r = \frac{dy}{d^2x} = \sqrt{2az-z^2}$, by substitution.

Cor. 1. When $z = o$, $r = o$, which shows that the evolute commences at the point A. When $z = a$, $r = z = 2$DC; consequently DC = CO′.

Cor. 2. The general expression for AK (page 143) is $\frac{dydx}{d^2x} - y$. Hence

AK $= \frac{2az-z^2}{2a}$, by substitution and reduction.

Now when $z = a$, $r =$ O′D = arc O′OA $= a$, and AP $= \frac{1}{2}a$, the radius of curvature being $\sqrt{2az-z^2}$, its square is $2az - z^2$. But $a^2 : 2az - z^2 :: \frac{a}{2} : \frac{2az-z^2}{2a}$, dividing the first and second terms by $2a$. Hence
AO′^{2} : AO^2 :: AP : AK, which is the property belonging to a cycloid whose curve is AO′, half its base O′P, and diameter of its generating circle AP.

Hence the evolute of the semi-cycloid AD is an equal semi-cycloid AO′.

(7.) As the preceding problem is of the greatest importance in demonstrating the properties of the pendulum, we shall give the pupil another investigation of it, the radius of curvature being expressed in terms of the radius of the generating circle and abscissa of the cycloid.

Let DC $= 2a$, or $a =$ the radius of the generating circle,

DF $= x$, FE $= y$, and the arc DH of the circle $= u$,

then $y = u +$ FH, by the property of the cycloid.

But FH $= \sqrt{2ax - x^2}$, by the property of the circle.

Hence $y = u + \sqrt{2ax - x^2}$, and $dy = du + d.(2ax - x^2)^{\frac{1}{2}}$.

But $du = \dfrac{adx}{\sqrt{2ax - x^2}}$, by eq. (1) p. 109. Therefore

$$dy = \frac{adx}{\sqrt{2ax - x^2}} + \frac{(a - x)dx}{\sqrt{2ax - x^2}} = dx\sqrt{\frac{2a - x}{x}}$$

$$= \sqrt{\frac{2a - x}{x}}, \text{ when } dx - 1.$$

Again,

$$d^2y = \frac{-a}{x\sqrt{2ax - x^2}}, \text{ by differentiating } \sqrt{\frac{2a - x}{x}}$$

$$\text{or } \frac{(2a - x)^{\frac{1}{2}}}{x}, \text{ by the rule for fractions.}$$

Hence, since the rad. of curv. $r = \dfrac{z^3}{-d^2y} = \dfrac{(1 + dy^2)^{\frac{3}{2}}}{-d^2y}$,

by substituting $1 + dy^2$ for its equal z^2, it follows that

$r = 2\sqrt{4a^2 - 2ax}$, by substitution.

COR. 1. Since the tangent at E is at right angles to OE and parallel to DH, GE is parallel and equal to HC. But HC $= \sqrt{4a^2 - 2ax}$, therefore OE $= 2$GE, and GE $=$ OG.

COR. 2. When $x = 0$, $r =$ arc AOO$' =$ O$'$D $= 4a = 2c$D, which is the radius of curvature at the point D. When $x = 2a$, $r = 0$, which shows that the evolute commences at A and terminates in O$'$.

COR. 3. Since AG $=$ arc CH $=$ arc AL, the arc AL $=$ OL. Also, the remaining arc LP $=$ the arc

DH = HE = CG; therefore the semicircle ALP = PO′. Hence the evolute AOO′ is a cycloid equal to the involute.

(8.) To find the area of the cycloid.

Let the origin of the space DRA (fig. p. 154) contained between the lines DR, RA, and the convex side of the curve be at the point D, and let an ordinate move from D along DT till it reach TE. Let DC $= 2a$, DT $= x$, TE $= y$, arc DH $= u$, and FH $= s$. Then $s^2 = 2ay - y^2$ by the property of the circle.

And $2sds = 2ady - 2ydy$, therefore $ds = \frac{ady - ydy.}{s}$

But $du - \frac{ady}{ds}$, by equation (1.) page 109.

Also $x = u + s$ by the equation of the cycloid.

Therefore $dx = du + ds = \frac{2ady - ydy}{\sqrt{2ay - y^2}}$, by substitution.

Consequently $ydx = \frac{2ay - y^2}{\sqrt{2ay - y^2}} \times dy = dy\sqrt{2ay - y^2}$

But ydx is the differential of the area DTE, and also of the area of the portion FDH of the circle. Hence

$\int(2ay - y^2)^{\frac{1}{2}}dy =$ area of DTE = area of FDH.

COR. When $y = a$, area of DRA = area of semicircle DHC. Hence area of DRA + area of DSB = area of the generating circle. But the area of the circle = circumference $\times \frac{1}{4}a$ = AB $\times \frac{1}{4}$AR.

Now, if from the area of the rectangle AB $\times$ AR, we take away the area of the spaces DSA, DSB, we leave

P

the area of the cycloid. Hence $AB \times AR - AB \times \frac{1}{4}AR = \frac{3}{4}AB \times AR$. That is, the area of the cycloid is $\frac{3}{4}$ of its circumscribing rectangle, and *three* times the area of the generating circle.

EXERCISES.

1. The wheel of a stage-coach is 5 feet high: required the length of the cycloidal arc AE (fig. first, p. 151) described by a nail in the rim, when the wheel has rolled over $\frac{1}{3}$ of its circumference?

2. If the tangent at E be continued to meet the axis produced; required the length of the subtangent, the numbers being as in the preceding example?

OBSERVATIONS ON THE PRECEDING PART.

The number of curves of the higher orders, whose properties have been investigated by different authors, is exceedingly numerous. We have the *Cissoid* of Diocles, the *Conchoid* of Nicomedes, the *Quadratrix* of Dinostrates, the *Logarithmic* curve and *Logarithmic spiral*, the *Cycloid*, *Epicycloid* and *Companion* to the cycloid, the *Paraboloid*, the *Catenary Curve*, the *Tratrix*, the *Magnetic Curve*, the *Curve of Sines*, the *Curve of Tangents*, the *Curve of Secants*, the *Equable Spiral*, the *Hyperbolic Spiral*, the *Logarithmic Spiral*, the *Lemniscata*, &c. &c. Though the properties of these curves be curious and interesting to the mathematician, their practical applications are extremely limited. The epicycloid has been proposed as the best form for the teeth of wheels and pinions, and the catenary as an *arch* of equilibration. The student who can spare time, and wishes to become acquainted with the properties of these curves, will find a full account of them, investigated without the *language* of the *Differential Calculus*, in Sir John Leslie's Geometry of Curves of the Second and Higher Orders. The two curves whose principal properties have been investigated in the preceding sections are the most important. The logarithmic curve is useful in exhibiting the law of the diminution of density of the atmosphere, and the cycloid in investigating the beautiful laws of the pendulum, and the descent of heavy bodies towards the centre of the earth.

QUESTIONS ON THE PRECEDING PART.

1. Since *no part* of an ellipse, parabola, &c. is a circle, how can they be said to have the same curvature as a circle of a certain radius?

2. What is meant by the term circle of curvature?

3. In determining the radius of curvature, must you differentiate oftener than once?

4. Describe the properties of the logarithmic curve?

5. How does it happen that the space contained between the ordinates, axis and curve, is *finite*, when its length is infinite?

6. What is that curve whose subtangent is *constant?*

7. Describe the properties of the cycloid, and illustrate them by the familiar example of a cart wheel?

8. If a circle revolve on the circumference of another circle, a point in its circumference will describe an epicycloid.

9. Describe an epicycloid by means of a penny piece and a halfpenny.

10. Illustrate by a similar example the nature of an involute and its evolute.

PROMISCUOUS EXERCISES, WITH OCCASIONAL HINTS FOR THEIR SOLUTIONS.

1. If x increase uniformly at the rate of 2, and y at the rate of 3, at what rate is the function $\frac{x}{x-y}$ increasing or diminishing when x becomes 10 at the moment y becomes 4?

2. If x increase uniformly at the rate of 1, and y diminish uniformly at the rate of 3, at what rate is the fraction $\frac{1}{x-y}$ increasing or diminishing when x becomes 10 at the same moment that y becomes 9?

3. What is the differential coefficient of

$$\sqrt{ax+bx^2+cx^3}?$$

4. If x increase uniformly at the rate of 1, what is the form and value of the expression which is increasing at the rate of $\frac{2+x}{2\sqrt{2x+x^2}}$ when $x = 10$?

5. It is required to inscribe a rectangle in a given circle whose diameter is 12 inches, so that the area of the rectangle multiplied by its longest side shall be a maximum?

6. What are the sides of the greatest rectangle which can be inscribed in a circle whose diameter is 10 feet?

7. Required the value of a fraction whose cube shall exceed its square by the greatest possible quantity?

8. There is a parabola whose abscissa is 9, and double ordinate 12; required the sides of the greatest possible rectangle which can be inscribed in it?

9. If x increase uniformly at the rate of 1, it is required to determine at what rate the value of the expression $a^3x^3-a^2x^2+ax+c$ is increasing when x becomes 10, the constant quantity a being 2? This is to be determined by the general principle of substituting $x+h$ for x in the above expression, finding the increment of the function, and then determining the value of the differential coefficient $\frac{du}{dx}$, the letter u being taken to represent the function. The pupil will now understand that the same process is to be employed when we speak of determining the differential coefficient *from first principles.*

10. Determine from first principles and also by a particular rule the differential coefficient $\frac{du}{dx}$ in the equation $u = \sqrt{ax^4-bx}$.

11. Determine by means of a certain rule the differential coefficient $\frac{du}{dx}$ in the equation $u = \frac{x}{\sqrt{a^2+x^2}}$.

12. What is the differential of the function axy^2z^3?

13. If x increase uniformly at the rate of 1, at what rate is the function $\sqrt{a^2-x^2}$ when x becomes 4, the constant quantity a being 5?

14. If x diminish uniformly at the rate of 1, at what rate is the function $\sqrt{ax-x^2}$ increasing or diminish-

ing when x becomes 10, the constant quantity a being 20?

15. Determine from first principles the differential coefficient $\frac{du}{dx}$ in the equation $u = (a+a^2x^2+a)^{\frac{3}{2}}$.

16. What is the value of u in the differential equation $du = \frac{x^{\frac{1}{2}n-1}dx}{a^n+x^n}$? As this differential does not seem to belong to any of the forms previously investigated, let us try what form it will assume by putting b^2 for a^n and y^2 for x^n. The numerator of the differential then becomes $\frac{2}{n}dy$, and the differential itself

$\frac{2}{n}\frac{dy}{b^2+y^2}$. But $\frac{dy}{b^2+y^2}$ is the form in (3) page 112.

$$\text{Hence} \int \frac{2}{n}\frac{dy}{b^2+y^2} = \frac{2}{n}\int\frac{dy}{b^2+y^2} = \frac{2}{nb^2}z,$$

in which b is the diameter of the circle, and z the length of the arc, y being the tangent of the arc.

17. Required the numerical value of the integral of $\frac{x^{\frac{1}{2}n-1}dx}{a^n+x^n}$ when x becomes 20000, n being 4, and a 10000?

18. Prove by the substitution in example 16 that

$$\int\frac{x^{\frac{1}{2}n-1}dx}{a^n-x^n} = \frac{1}{nb} \times \log.\frac{b+y}{b-y}.$$

19. Prove, by means of the same substitution, that

$$\int\frac{x^{\frac{1}{2}n-1}dx}{\sqrt{a^n+x^n}} = \frac{2}{n}\log.(y+\sqrt{b^2+y^2}).$$

20. Demonstrate by the same substitution, that $\int \frac{x^{\frac{1}{2}n-1}dx}{\sqrt{a^n - x^n}}$ is equal to the arc of a circle, whose sine is y, and radius b, when multiplied by $\frac{2n}{nb}$.

SYNOPSIS OF THE PRINCIPAL SIGNS EMPLOYED IN THIS BRANCH OF SCIENCE.

1. The sign $\propto$ is used to denote that the quantities between which it is placed *vary* as each other. Thus, if s denote the space passed over by a body moving with the velocity v during the time t, then $s \propto tv$, denotes that the spaces passed over by bodies moving with different velocities and in different times are as the product of the times and velocities.

2. The sign ∞ denotes *infinity*. Thus $\frac{1}{0} = \infty$ denotes that the nearer the divisor approaches to 0, the nearer the quotient approaches to infinity.

3. The letter d placed before a single letter, or a point above it, denotes the rate at which the variable represented by it, varies; that is, its differential or fluxion. Thus dx or $\dot{x}$ denotes the rate at which the variable quantity represented by x is increasing. When placed before a compound expression it denotes the rate at which the value of the compound expression is increasing when the independent variable or variables are passing certain limits. Thus $d.(ax^2 + x^3)$ denotes the rate of increase of $ax^2 + x^3$ or its differential.

4. The letters F, f, ϕ, ψ, are employed instead of the term *function*. Thus Fx may denote any expression, as $(a+x)^2$, $\sqrt{x}$, $\sin x$, &c. into which x enters as the independent variable. $f(x, y,)$ may denote axy, $\frac{x}{y}$, $ay+bx^2$, &c. or any expression into which x and y enter as independent variables.

5. The letter $\int$ denotes the integral or expression from which a given differential has been derived. Thus $\int nx^{n-1}dx = x^n + c$.

6. The Greek letter Δ denotes the difference between two successive states of an independent variable or function. Thus, if x receive the increment h, and become $x+h$, then $\Delta x = h$. Also $\Delta(x^3) - 3x^2\Delta x + 3x\Delta x^2 + \Delta x^3$.

7. The sign Σ denotes the function or integral from which a given difference has been derived. Thus,

$$\Sigma\,(3x^2\Delta x + 3x\Delta x^2 + \Delta x^3) = x^3.$$

The sign d bears the same relation to Δ that $\int$ does to Σ.

8. The sign $\int_a^b$ denotes that the value of the integral is to be found for two particular values, a, b, of the independent variable, and the difference of these integrals to be taken. Thus, $\int_a^b 2xdx = a^2 - b^2 - 5$, when $a = 3$ and $b = 2$.

The sign $\int_0^\infty$ denotes the value of the integral between the limits $x = 0$ and $x =$ infinity.

9. The sign $\overset{-1}{\text{F}}$, $\overset{-1}{\phi}$, denotes an *inverse* function.

Thus, if $\mathrm{F}x$ denote $\sin x$, then $\mathrm{F}^{-1}x$ or $\sin^{-1}x$ denotes the *arc* of which x is the *sine*.

10. The sign $d_x y$, which has lately been used to a considerable extent by the Cambridge mathematicians, denotes "the differential coefficient of y taken with respect to x," and signifies the same thing as $\frac{dy}{dx}$ Also $d_x^2 y$ is used instead of the more elegant form $\frac{d^2y}{dx^2}$.

11. The sign $\int_x$ is frequently employed by Cambridge writers, and denotes the integral, x being the independent variable. Thus,

$$\int_x nx^{n-1} \text{ denotes} \int nx^{n-1}dx.$$

I entirely agree with Mr. Woolhouse in thinking that these signs have no recommendation whatever over those already established and incorporated in all the most valuable works of science. I am strongly of opinion that every change of notation which does not possess *decided* advantages over that universally used, has a direct tendency to bewilder the pupil and retard the progress of *real* science.

APPENDIX.

CONTAINING A CONCISE VIEW OF THE RISE AND PROGRESS OF THIS PART OF MATHEMATICAL SCIENCE.

(1.) THE earliest geometry of which we have any account consisted in the comparison of figures bounded by straight lines. Mathematicians, either prompted by curiosity or impelled by necessity, were, however, soon found engaged in the solution of a more difficult class of problems, viz. in the comparison of figures bounded by *curved* lines with those bounded by *straight* lines. The earliest problem of this kind which engaged the attention of geometers was the determination of the ratio of the areas of circles of different diameters. By inscribing polygons having the same number of sides in two given circles, they showed that these polygons were to each other as the squares of the diameters of the circles. The same ratio was also shown to exist between similar circumscribing polygons. By continually doubling the number of the sides of the inscribed and circumscribing polygons, they were shown to approach nearer and nearer to equality, and consequently to the intermediate circle.

Hence it is obvious, that the circles which are the limits of these polygons, must be to each other in the same ratio. This process is called the *Method of Exhaustions.*

(2.) The curvilinear surfaces, which came next to be compared with those bounded by straight lines, were the *Conic Sections.* Now, as regular polygons could not be inscribed in these figures, geometers were under the necessity of inventing another mode of comparison, which may be viewed as the next decisive step in a direct line to the discovery of the Fluxionary or Differential Calculus. This method consisted in inscribing and circumscribing an indefinite number of rectangles, the breadths of which being diminished *ad infinitum,* their *sum* approached to equality and to the intermediate curvilinear figure.

(3.) Archimedes, the greatest geometer of antiquity, invented another method of comparing the surfaces of figures bounded by curves with those bounded by straight lines, which may be viewed as the origin of the method of *Indivisibles,* which we shall afterwards consider. By comparing the sum of a series of numbers diminishing by equal differences with the greatest term, and also the sum of the squares of these numbers with the square of the first, Archimedes succeeded in determining the areas of certain curvilinear spaces and the solidities of solids of revolution.

(4.) The demonstrations of the ancients, though free from every objection with regard to accuracy, were so extremely tedious, that other methods were eagerly sought for by which the process might be greatly abridged. Cavallerius, an Italian mathematician, proposed the method of ***Indivisibles***, which is sometimes called the method of Cavallerius. The same principle was further extended by Dr. Wallis, under the title of the "Arithmetic of Infinites."*

As this method was extensively employed in determining the areas of surfaces and the capacities of solids, and is still frequently employed for the same purpose, it will be useful to give the learner a short account of the principles on which it is founded. According to this method, lines are made up of indefinitely small points, surfaces of an indefinite number of parallel lines, the breadths of which are considered as indefinitely small, and solids of an indefinite number of parallel planes, or rather indefinitely thin plates or laminæ. If the pupil take a piece of cloth formed of very fine threads, and cut it into any figure, a triangle for example, keeping the threads parallel to the base; and if he conceive these threads diminished in thickness, and increased in number, he will obtain a clear view of the mode by which surfaces are supposed to be made up. A book supposed to have the leaves indefinitely thin, affords a simple illustration of the manner in which solids are conceived to be formed.

Before applying this method, the pupil must understand the rules for finding the sums of the following series.

* "L'Arithmétique des Infinis est de l'invention de Wallis, célèbre Anglois, qui, malgré son antipathie pour la nation Françoise, mérite bien qu'on lui rende l'éloge qui lui est dû." —Deidier. La Mesure des Surfaces. Preface, xi.

1. The sum of the series $0+a+2a+3a+4a+$ &c. to na is equal to half the sum of the first and last multiplied by ~~half~~ the number of terms, that is, $\frac{na}{2} \times (n+1.)$ If the number of terms be *indefinitely* great, then the limit of the sum is $na \times \frac{n}{2}$, or the last term multiplied by half ~~the~~ number of terms.

2. The sum of the series $0+a^2+(2a)^2+(3a)^2+$ &c. to $(na)^2$ is equal to $\frac{1}{3}n \times (na)^2$ when n is infinite.*

EXAMPLES.

1. To find the area of a triangle.

Let the triangle be supposed to be made up of lines parallel to the base. Since the lines commence at the base, and terminate in the vertex, the greatest term is equal to the base, the last to 0, and the number of terms equal to the perpendicular. Hence the area is equal to the base multiplied by half the perpendicular.†

2. To find the solidity of a square pyramid.

The pyramid being supposed made up of an indefinite number of thin plates parallel to the base, these plates increase in magnitude according to the series 0, a^2, $(2a)^2$, $(3a^2)$, &c. to the square of the side of the base. Hence their sum is $\frac{1}{3}n \times$ area of the base. But $n =$ the perpendicular height, hence the solidity of the pyramid is found by multiplying the area of the base by one-third of the height.

* Emerson's Arithmetic of Infinites.

† Principles of Geometry, page 128.

These examples will be sufficient for giving the learner an idea of the application of the Arithmetic of infinites, or the method of indivisibles, to determining the areas of surfaces, and the capacities of solids. Those who wish to obtain a complete view of this curious subject, may consult Emerson's Arithmetic of Infinites, or the more extensive work of Deidier, entitled " La Mesure des Surfaces et des Solides."

(5.) The first person who considered certain quantities as formed or generated by continued motion, seems to have been Baron Napier, the celebrated inventor of logarithms. According to the views of Napier, two points were supposed to move along two straight lines, the one with a *uniform* motion, and the other with a *variable* motion continually accelerated; the lines generated representing the logarithms and their corresponding numbers. The ideas of Napier concerning the mode of generating numbers and their logarithms, are so like those of Newton in his manner of viewing the fluxions of quantities, that it is exceedingly probable the views of Napier may have turned the thoughts of Newton into the tract which terminated in the most splendid discovery of modern times.*

(6.) Dr. Barrow took a more extensive view of the formation of all kinds of magnitude by motion than Napier had done, and seems to have arrived almost at the point from which he might have obtained a view of the fertile regions which the penetrating eye of Newton was first destined to behold.

* Montucla. Histoire des Mathématiques, tom. ii. pages 16—97.

(7.) According to Newton, all quantities are generated by motion. A line by the motion of a point, a surface by the motion of a line either constant or variable, and a solid by the plane either retaining the same magnitude or varying according to a certain law. All other quantities, in whatever way they may be expressed, are conceived to be generated in a similar manner. In this way he obtained the ratio of the velocities or of the fluxions of quantities, and by the inverse process their *fluents* or integrals. In the preceding pages we have chiefly adopted the views of Newton as extended and modified by Maclaurin and d'Alembert.

(8.) About the same period, and apparently without knowing what Newton had done, the celebrated Leibnitz invented the Differential and Integral Calculus, which corresponds to the direct and inverse method of fluxions. According to the views of Leibnitz, quantities are composed of other quantities diminishing in value with almost infinite rapidity, these latter quantities being termed *Infinitesimals* of different orders. Let $(x+h)^n$ be expanded as follows:

$$(x+h)^n = x^n + nx^{n-1}h + n.\frac{n-1}{2}x^{n-2}h^2 + \&c.$$

Now, since $x : h :: h : h^2 \cdot\cdot h^2 : h^3$, &c. it follows that if h be indefinitely small compared with x, h^2 will be indefinitely small compared with h, h^3 indefinitely small compared with h^2, &c. Hence, since the successive terms in the preceding or any similar developement are multiplied by h, h^2, h^3, &c. these terms will go on diminishing in such a manner, that each term will be indefinitely small compared with that which precedes it.

If these infinitesimals be neglected in the developement of the *difference* between the two successive states of the function, we shall obviously arrive at the same result as taking the *limit* of that difference. The methods of Leibnitz and Newton, therefore, differ only in the notation and the peculiar modes of viewing the subject, or what is commonly called the *Metaphysics* of the Calculus.

(9.) We have already noticed the mode of viewing the subject adopted by La Grange, in his profound works entitled "Le Calcul des Fonctions" and "Théorie des Fonctions Analytiques," which, though full of interesting matter, are not sufficiently elementary for communicating the first principles of the science to youth.

(10.) Though I have only mentioned the names of those whose writings had the most direct tendency towards the discovery of the principles of the calculus, the names of Roberval, Fermat, and above all, Des Cartes, may be mentioned with distinction. After the discoveries of Newton and Leibnitz, those of Euler are second only as far as the first principle is concerned, but are no less important in extending the limits of the science and in giving form and consistency to the whole. The labours of the Bernoullis had also a powerful influence on the progress of this branch of Analysis. In our own country the writings of Taylor, Maclaurin, Emerson, and Simpson, may be mentioned as those which had the greatest effect in reducing the Fluxionary Calculus to a regular form and diffusing a taste for the science among English mathematicians. But I must close these remarks by referring the inquisitive student for farther information to Montucla's "Histoire des Mathématiques," or to the Preface to La Croix's large work, entitled "Traité du Calcul Différentiel et Intégral."

THE END.

London:
Printed by Samuel Bentley, Dorset Street, Fleet Street.